David Beer

Metric Power

palgrave
macmillan

David Beer
Department of Sociology
University of York
York, United Kingdom

ISBN 978-1-137-55648-6   ISBN 978-1-137-55649-3   (eBook)
DOI 10.1057/978-1-137-55649-3

Library of Congress Control Number: 2016944241

Cover illustration: © Enigma / Alamy Stock Photo

Printed on acid-free paper

This Palgrave Macmillan imprint is published by Springer Nature
The registered company is Macmillan Publishers Ltd. London

For Mum and Dad
Some things can't be measured.

Preface

All books inevitably carry some of the flavour of their times. The pages of this text are no different; they have undoubtedly been embossed with the moment in which it was produced. This book emerged during a time when the presence of metrics was notably escalating in higher education. One recent report has described this as a kind of 'metric tide' (Wilsdon et al. 2015). I suspect that my ideas have been tinged by the tincture of these apparent changes in academia. Yet this is certainly not a book about academic work or higher education. Rather, this is a book that I hope will speak to people with a general interest in how power operates today. It is most obviously a book about data assemblages, culture, and new media forms, but I hope that it will also be of some use to those with an interest in questions of power, governance, cultural politics, and political sociology. Amongst these more general aims, and to give an opening feel of its content, this book attempts to provide the reader with the conceptual means for thinking critically about the role of metrics in contemporary society and culture. The book is not comprehensive in its descriptions of the types of metrics that act upon us, but it is hoped that the ideas contained here can be applied widely in response to the powerful use of metrics in the ordering and governance of our lives. I will explain this in far more detail in Chap. 1; I would like though to use this very brief preface just to offer some reflections on the cultivation of the ideas contained in this book and the approach that I have taken.

Sometime during the academic year 2011–2012, I devised a new post-graduate module. I gave that module the rather provisional-feeling title 'Digital By-Product Data and the Social Sciences'. It was a little wordy, but I wasn't sure that the label 'big data' was appropriate at that time. I knew that the title was just a vehicle for capturing something that was unfolding in social science. As I began to write this book, I was delivering this module for the third and final time (it is to be replaced by a new incarnation in early 2016). Each year the module had just over 30 students on it—I thank the students for their enthusiasm and depth of discussion over the last three years. Working on this module and working with these students has really helped to reveal the potential gaps that need attention in this emergent field of research. Initially this module was intended to provide a space in which students could begin to think about how they might use new types of digital data for doing social research. As I taught the module, things seemed to change. I realised that in order for these general objectives to be achieved we needed to do more to enhance our critical and imaginative faculties, particularly when presented with these emergent forms of data. The data itself was not necessarily the problem, although data access was always likely to be a pre-occupation; the difficultly was instead in finding the means to think critically about them. The problem was in finding ways to craft questions that might be asked about, through and with such data. My conclusion was that we need to see these big data differently. We need to foster some alternative perspectives that go beyond those scripted into the data. We need to look at them with fresh eyes. We need to carve out some new vantage points that will allow us to see what types of questions might be asked with such data, to see how these data become part of the social world and to see how we might respond critically to the ways that these data shape and cajole the social world into new formations (or maintain obdurate social orders). In short, we needed to work on being more assertive in our response to the emergence of big data. The stuff that is called 'big data' undoubtedly creates important questions about our analyses and techniques, but they create more pressing questions about the sharpness of our imaginations. As I argue later in this book, the challenge of big data is as much one for the imagination as it is for our technical skills—it is a challenge of thoughtfulness, not just of learned skill or know-how. It is in the provocation of the imagination that this book intends to make an intervention in debates on the new types of data and what they mean.

As I taught the module, it became clear to me that we needed more engagement with the politics of the data themselves, and in particular we needed to see how data, in the form of metrics, could be seen to be measuring us in new and powerful ways. As a result, it seemed important to explore the relations between metrics and power. The results of the insights that I accumulated via this module are to be found permeating through the pages of this book. My suggestion is that by developing such critical vistas we may see how to utilise new forms of 'big data' and how we might reconceive our research questions. More importantly though, we might also then come to understand the part that metrics play in the ordering of the social world and in the shaping of our lives. Any analysis of big data should start from such a vantage point. We need to understand how metrics implicate and are implicated by the versions of the social that they purport to reveal. We also need to understand our own participation in both revealing and potentially challenging the measures of the world that they produce.

When I was a good way through the background work for this book, somewhere around the mid-point in the writing process, I stumbled upon an interview with Michel Foucault. Foucault's words jumped out of the page; they just seemed to chime with the work I was doing for the book. In the interview, Foucault (1991a: 73–74) describes his own work in the following terms:

> My work takes place between unfinished abutments and anticipatory strings of dots. I like to open up a space of research, try it out, and then if it doesn't work, try again somewhere else. On many points…I am still working and don't yet know whether I am going to get anywhere. What I say ought to be taken as 'propositions', 'game openings' where those who are interested are invited to join in.

Whilst writing I found the open-ended and exploratory sentiment of this passage resonated. It spoke directly to the type of project that I was trying to develop—a project based on an attempt to join together some disparate dots. Clearly, this book is not able to explore all of the permutations and intersections between metrics and power; this is a massive project that will need continuous and close attention. I hope though that this book will be seen, in Foucault's words, as a set of propositions and openings. My hope is that the conceptual materials I develop here will form a framework for further and more nuanced analyses of the relations between metrics and power, a set of relations that needs renewed attention in the current context (see Chap. 1). This book then, to try to absorb some of Foucault's style and sentiment, is a kind of invitation to the reader to join in. The book is aimed at helping us to work together to join some of the dots that pattern around the social implications of a rising interest in the capabilities and possibilities of metrics, numbers, and calculation. My suggestion, for the moment at least, is that we begin to sketch these connections in pencil.

References

Foucault, M. (1991a). Questions of method. In G. Burchill, C. Gordon, & Miller, P. (Eds.), *The Foucault effect* (pp. 73–86). Chicago: The University of Chicago Press.

Wilsdon, J., Allen, L., Belfiore, E., Campbell, P., Curry, S., Hill, S. et al. (2015). *The metric tide: Report of the independent review of the role of metrics in research assessment and management*. doi: 10.13140.RG.2.1.4929.1363.

David Beer
York, UK

Acknowledgements

Thanks go to a number of my colleagues for discussing and encouraging the ideas captured here. Particular gratitude goes to Gareth Millington, Ruth Penfold-Mounce, Paul Johnson, Ellen Annandale, Daryl Martin, Les Back, Nathan Manning, Brian Loader, Alex Hall, Rowland Atkinson, Andrew Webster, and Nisha Kapoor. Additional thanks go to Rowland Atkinson, who invited me to give a short ten-minute presentation at the *ISRF Workshop: Critique and Critiques* which took place in York on 13 May 2014; that presentation eventually transformed into this book. Thanks are due to the audience at that event for their helpful and enthusiastic questions. Huge gratitude goes to Helen Kennedy, who acted as a reviewer for the whole manuscript. Her careful and thorough comments undoubtedly helped me to strengthen the book. I thank her for the time and thought she put into her review.

The content of the book is entirely new, but a handful of paragraphs from three short pieces have made their way, in revised form, into the final manuscript. Thanks go to Mark Carrigan for allowing me to publish a short piece on the Apple Watch on his *Sociological Imagination* blog, which helped me to develop a short passage that is included in Chap. 1. A fragment of Chap. 3 is based upon a few passages drawn from the short magazine piece 'The New Circulations of Culture', which was published in the magazine *Berfrois*—thanks go to Russell Bennetts who edits *Berfrois*. Finally, thanks also go to Nathan Manning for the invitation to write a review of Btihaj Ajana's book *Governing Through Biometrics* for the journal *Information, Communication and Society*. Writing that review really helped me to formulate the ideas in the section on biometrics contained in Chap. 2.

This book is dedicated to my loving mum and dad. I'd also like to mention Nona, who has been a great source of support and humour along the way. As always, I give special and immeasurable thanks to Erik and Martha.

Contents

1

Introducing Metric Power

Without wanting to open this book with a cliché, I feel compelled to admit that I once worked in a panopticon of sorts. It was a call centre. At the time, which was around the mid-to-late 1990s, it was one of the first of a new raft of call centres that were popping up in various destinations in the UK and around the world. Many organisations at the time were moving to phone-based service provision. Help lines and online customer services were the watchwords of an expanding service-based economy. Foucault, I suspect, would have had a field day. The cavernous hangar-style building had no windows (except for those located in the small dedicated 'break-out' areas). There was a 'centre desk' around which the call centre was organised, with desks orbiting out from that central point. The office was open plan and the people on the centre desk were charged with monitoring and surveilling the 'operatives' answering the phones. Those on centre desk could see everyone in the room; plus they could also see the individual operatives' metrics in real time on the dashboards projected on their desktop monitors. At the beginning of my shift, I logged onto both the phone system and a desktop computer—two interconnected technologies of surveillance through which all of my daily tasks were routed. Working changeable shifts we had to 'hot-desk'.

© The Editor(s) (if applicable) and The Author(s) 2016
D. Beer, *Metric Power*, DOI 10.1057/978-1-137-55649-3_1

It was an environment of impermanency and mobility; desks changed, staff turned-over, calls kept flowing.

The computer system tracked our use of the front-end software designed for dealing with customer queries. Yet it was the phone system that was far more powerful in gathering the most important metrics about our labours. This system allowed the centre desk to watch what each person was doing. Each action was made visible in the metrics—were they on the phone, were they wrapping up a previous call, were they on a break, were they in the toilet. Each of these activities was recorded by the phone system. The metrics were watched in real time, and weekly reports were also produced for performance management purposes. There were some minor forms of resistance, but the system was rigid and allowed little manoeuvre.

It is perhaps no wonder that I can now see this experience through the lens of Foucault's writings. This was a time spent with power operating through observational knowledge—in which, as Foucault once put it in a lecture delivered in Rio de Janeiro, 'knowledge about individuals…stems from the observation and classification of those individuals, from record- ing and analysing their actions, from their comparison' (Foucault 2002b: 84). How many calls, how long you'd taken on your breaks, the duration of your calls, how long you take to get to the next call, how long you've taken in the toilet that week, and so on. These metrics were a central part of the management of the people working there. And this is before we add the extra dimension of surveillance that came with the covert listening of calls by the anonymous group of people simply referred to as 'compliance'. You didn't know when you were being watched, listened to, or measured.

But that was over 15 years ago and things have changed. The scale of metrics in everyday life has only amplified to a volume that couldn't have been foreseen back then. As Ronald E. Day (2014: 132) has recently observed, 'no longer is surveillance of the individual enough, but now he or she is co-located within predictive matrixes of actions and objects through linked associations with other subjects, objects, and events in databases and their indexes'. Indeed, as I hope will become clear in this book, what I will call *metric power* is not simply concerned with surveil- lance—surveillance is important, but *metric power* is not limited to the art of watching, nor can it simply be reduced to the internalisation of the feeling that we might be watched. What this call centre example illustrates, despite this amplification of metrics in recent years, is that metrics have had an ordering role in the social world for quite some time. This book keeps an eye on this history, going back beyond this short 15-year period, whilst also attempting to think about the intensifying role of metrics in various aspects of our everyday lives and the social world today. Stiglitz et al. (2010: xvii) recently surmised that in 'an increasingly performance-oriented society, metrics matter'. They matter, they argue, because 'what we measure affects what we do'. This is a fairly obvious conclusion perhaps, but it is one that we should nevertheless continue to concern ourselves with.

The types of ramping up of metrics that I'm alluding to when recalling this call-centre experience has led Will Davies (2015a: 222), referring to Jeremy Bentham and the behaviourist expert John B. Watson, to conclude that:

> The combination of big data, the narcissistic sharing of private feelings and thoughts, and more emotionally intelligent computers opens up possibilities for psychological tracking that Bentham and Watson could never have dreamed of. Add in smartphones and you have an extraordinary apparatus of data gathering, the like of which was previously only plausible within university laboratories or particularly high-surveillance institutions.

Thus, he powerfully infers, we are 'living in the lab'. The crucial point that we can extract from this is that the very apparatus of measurement has drastically expanded; this expansion is allied with a set of cultural changes in which the pursuit of measurement is seen to be highly desirable.

Within the vast circulating swirls of data that have become so powerful, we can see metrics as being those data that are used to provide some sort of measure of the world. In this book, I will suggest that we are created and recreated by metrics; we live through them, with them, and within them. Metrics facilitate the making and remaking of judgements about us, the judgements we make of ourselves and the consequences of those judgements as they are felt and experienced in our lives. We play with metrics and we are more often played by them. Metrics are a complex and prominent component of the social as they come to act on us and as we act according to their rules, boundaries, and limits. Metrics are a deeply woven aspect of everyday lives and the social world in which these lives are conducted. Metrics are a prominent and powerful part of the governance of contemporary life: from smartphone apps that measure our sleep and exercise, to the data produced by our transactions or social media profiles, through to the measurement of our performance at work, our health, and the financial systems of the global economy. We even have smart meters that help us to manage our household energy consumption and an emergent industry around the measurement and manipulation of something as apparently immeasurable as our emotions—with metrics on well-being used in attempts to maximise our output and efficiency (see Davies 2015a; Brown 2015b). An example of this kind of measurement of emotions can be found in the services offered by the-happiness-index.com, who provide customised assessment of happiness and well-being in order for companies to maximise their employee engagement and productivity. It would seem that metrics, as we will discover, often have a purpose: they are laced with intentions.

In short, metrics are now an embedded, multi-scalar, and active component of our everyday lives—they are central to how those lives are ordered, governed, crafted, and defined. With all of this in mind, it could even be claimed that systems of measurement are at the heart of the very functioning of the social world as it is today—but perhaps we are getting ahead of ourselves. Metrics themselves are nothing new; these systems have a long history. Populations have been measured in various ways for a long time (see e.g. Foucault 2007). As Foucault (2013: 134) has pointed out, 'we should not forget that before being inscribed in Western consciousness as the principle of quantification...Greek measurement was an immense social and polymorphous practice of assessment, quantification, establishing equivalences, and the search for appropriate proportions and distributions' (this periodisation is also discussed in relation to the transformation brought on by written numerals in Kittler 2006). I will discuss this long history in detail in Chap. 2. Yet there is little doubt that these systems of measurement have escalated and intensified over recent years, especially with the rise of new data assemblages and their integration into the very fabric of our lives (see Beer 2013). As Espeland and Sauder

[handwritten margin note: metrics are central to our lives]

(2007: 1) put it, 'in the past two decades demands for accountability, transparency, and efficiency have prompted a flood of social measures designed to evaluate the performances of individuals and organization'. Newspaper accounts of extreme examples of the data-dense working environment reveal how a 'staggering array of metrics' can be used to hold people accountable in minute detail (see the reports in Kantor and Streitfeld 2015). And then we also have major quantification-based reports such as the 'Global Human Capital Trends 2015' report (Deloitte 2015), which the authors describe as 'one of the largest longitudinal studies of talent, leadership and HR challenges'. That particular report examines different aspects of 'human capital' and provides quantifications of leadership, engagement, contingent workforce use, performance management, and so on. Alongside these quantifications, the report pushes for new approaches to quantifying 'human capital' by 'reinventing HR' and taking advantage of the new 'people data' and 'people analytics'. The report's message is that the performance of people and organisations can be quantified in multiple ways, with yet plenty of new ways in which we should be furthering this quantification. The ramping up of metrics is depicted as the only sensible and desirable future. Just to pick out one phrase by way of illustration, this report informs us that 'high potential young employees want regular feedback and career progression advice, not just "once and done" reviews' (Deloitte 2015: 53). The push then, in this report, is towards *ongoing* and *increasingly granular* metric-based evaluations and judgements. Espeland and Sauder's observation, which we should note was written before smartphones and social media really took hold, seems highly pertinent then. As they claim, despite this long history 'what is relatively new is the public nature of social statistics' (Espeland and Sauder 2007: 4). Metrics have become embedded and so has their authority, to the point where, Espeland and Sauder (2007: 5) claim, we have 'trouble imagining other forms of coordination and discipline or other means of creating transparency and accountability'. It is difficult to imagine a world that is not ordered by metrics or defined by the prominence of the desire to metricise everything.

In this context, this book will explore the social role, significance, and consequences of metrics and data. More specifically, it aims to examine the linkages between metrics and power in the contemporary setting. It is commonly suggested, not just by academics but also in the popular media, that many of the most significant technological developments of our age will centre around data and metrics. Given its potential importance and apparently increasing role, the book will look at how measurement links into power, governance, and control. In order to do this, the book will focus upon one particular set of relations: these are the relations that exist between *measurement, circulation,* and *possibility.* It is this set of connections that forms the central focus of this book. The reason that I focus upon these relations is because they allow us to see the politics of what is measured whilst also considering this in concert with the way that those measures move out into the world. It is crucial, I argue here, to understand these relations if we are to grasp the various ways in which metrics interweave into power structures. It is argued that it is in these relations that we can locate and understand what might be thought of as *metric power.* As such, the concept of metric power is focussed upon unravelling the power dynamics that underpin or reside within big data and other related phenomena.

Contextualising *Metric* Power

Before developing this central concept further, let us first reflect a little more on the context in which these relations operate. To take just one prominent example, the recently launched Apple Watch is perhaps emblematic of both our creeping connectivity and the extension of metric power into our lives. Metric power is not just about such hi-tech devices, but these provide us with a visible marker how our bodies can be directly interfaced into the infrastructures of metric harvesting. These are devices from which metrics are drawn and then used to provoke and stimulate responses. When you look at the marketing that has accompanied the launch of the Apple Watch, you actually find that this kind of bodily and nervous connectivity is a central part of how the watch is being sold (for access to the marketing materials discussed here, see Beer 2015a). We are told that it will provide a more 'haptic' experience. This is a tactile and sensory set of connections, with the watch sharing sensory information with the body. It extracts information such as heart rate, using its sensors placed on the skin, whilst buzzing with notifications and bodily interventions.

The device is presented as being part of a lifestyle in which our connectivity becomes the means of self-improvement and heightened experience. As the marketing video dedicated to the topic of health and fitness suggests, this watch 'gets to know you the way a good personal trainer would' (again, see Beer 2015a). That is to say that it takes on an active role in guiding your lifestyle by suggesting goals and activities. The promise is that you will become less sedentary and the watch's presence on your body will stimulate and provoke action. These devices—like the Apple Watch and others like Fitbit and Jawbone—are not just about personal data and its use to shape lifestyle though. Insurance companies now offer reduced premiums and other incentives if you are prepared to share the metrics from your wearable device with them (see e.g. Stables 2015). The lifestyles captured in those metrics become part of the decisions made about people. And as we work through this book, we might also wonder how the sharing of such lifestyle metrics will shape the behaviours that feed into those metrics. As I have argued elsewhere, these are 'productive measures' (see Beer 2015b), they produce outcomes as well as measuring them—this is what Espeland and Sauder (2007) refer to as 'reactivity'. This all chimes with Nikolas Rose's (1991: 691) much earlier observation that neoliberal rationalities 'require a numericized environment in which these free, choosing actors may govern themselves by numbers'. Perhaps then the very growth of systems of measurement that we have seen is a kind of marker of neoliberal rationalities at work—found in the desire to measure. Wearable devices like the Apple Watch make it possible to govern ourselves by numbers in much more nuanced, personalised, and direct ways.

Yet we should remind ourselves that the Apple Watch and other wearables like the Jawbone—which captures and sets targets for calorie burn, number of steps, heart rate, and the like—are just one of the more visible tips of a large iceberg that has formed as the infrastructures melt and freeze with cultural tropes that promote the social position of metrics. *Metric power* can appear in far more mundane and crude forms than these sparkling, sleek, and flashy smart devices would suggest. Ranging across our consumption, our work, and our leisure, metrics play a powerful role.

how

M

C

P

World

↓

+
power
over
lives

Responding to this context, this book will explore a range of conceptual and empirical resources for understanding the relations between *measurement, circulation,* and *possibility.* Its explorations will not just centre on these visible hi-tech gadgets, it will also be concerned with these relations as they exist in much more humdrum forms. However, if we pause just to a reflect on the sharing of wearable device metrics with insurance companies by way of a short illustration, we might conclude that the way that people are *measured* in their lifestyle and exercise choices *circulates* through the commercial archives of the insurance company which then shapes the *possibilities* for the treatment of that individual. This will also then have implications for how people live—as they come to live a predictive life in which they adapt to the measures. This is just one example, but it suggests that these are complex and recombinant processes that illustrate, from the outset, the types of relations that this book attempts to explore. With this in mind—and with this focus on the relations between measurement, circulation, and possibility at its centre—this book is aimed at contributing towards expanding our understanding of the role of metrics in the performance of contemporary society. The implicit argument of the book is that before we see if these brand new and expanding types of metrics give social scientists a useful 'measure of the world', a measure that they can use in their own analytics, first we need a more conceptual, contextual, and politically sensitive appreciation of metrics and data. As I argued in the conclusion to my previous book on data circulations, social scientists need to be at the forefront of the critique of new types of digital or 'big' data as well as being involved in using that data (see Beer 2013: 172–174). As such, and as I have already intimated, this book asks how we might understand and conceptualise the power dynamics behind the so-called big data (for a particularly useful definition of this elusive term, see Kitchin 2014a: 67–79).

When it comes to quantification, Espeland and Stevens (2008: 402) point out that 'sociologists have generally been reluctant to investigate it as a sociological phenomenon in its own right'. They claim that the focus has tended to be upon 'the accuracy of the measures' rather than their 'social implications'. Despite the presence of big data as both a set of phenomena and a prevalent concept, and despite the general accumulation of new types of metrics, Espeland and Stevens' point still pertains (see also Rottenburg and Merry 2015: 3). In short, we perhaps surprisingly still have much to do to develop a sociological understanding of the social implications and roles of metrics (despite some important and notable works, which will be discussed in detail throughout this book).

Helpfully, Rob Kitchin (2014a) has recently outlined an agenda for a critical encounter with the 'big data revolution'. The position taken by Kitchin (2014a: 184) is that there 'needs to be a more critical and philosophical engagement with data'. Elsewhere danah boyd and Kate Crawford (2012) have suggested that big data can be misleading if taken out of context. Keeping these two parallel arguments in mind, this book sets out to provide this kind of critical, conceptual, and contextual encounter with big data. This is not a book about big data as such, it is intended to be about something broader with regard to the power of metrics, but it nevertheless tackles these metrics in the context of big data. The book will focus frequently on big data because this is a crucial way in which we have come to understand the current intensification of measurement. Much of what I say in this book can be seen as a part of these debates around big data, yet it aims to offer a more detailed engagement with how power might be conceptualised in such a setting. One way to do this, as I will discuss in a moment, is to try to situate debates on metrics and big data within some broader streams of work and debates around political economy and governance. These help us to see the cultural politics that usher in the more technical and infrastructural aspects of measurement.

Kitchin (2014a: 185) states that 'given the utility and the value of the data there is a critical need to engage with them from a philosophical and conceptual point of view'. The added notion of value is interesting here. The role that data play, often in the form of metrics, is to provide a utility around which value can be imagined or extracted. Data and metrics have come to underpin much of the value generation of contemporary capitalism, as Nigel Thrift (2005) suggested around a decade ago—which has also recently been restated in the form of 'digital capitalism' and the 'digital depression' of the economic crisis by Dan Schiller (2014). As Kitchin suggests, it seems there is a pressing need to conceptually engage with the relations between data and value, and it is through metrics that this can be elaborated and explored. Metrics are a form of data through which value can be measured, captured, or even generated. Metrics are the means by which data can be used to ascertain value.

In Chap. 2, I will place this apparent data revolution within a historical context, but there is a sense nevertheless that this is something that is unfolding rapidly and which is substantively different from the social formations of the past. I will question this a little in the following chapters, but Kitchin's claims are well worth reflection at this opening juncture. His claim is that:

> The data revolution is in its infancy, but is unfolding quickly. In just a handful of years open data, data infrastructures and big data have had a substantial impact on the landscape, accompanied by vocal boosterist discourses declaring their positive disruptive effects. (Kitchin 2014a: 192)

Kitchin takes a considered and critical approach to such an unfolding data revolution, but we can see that he notes a rapidly emergent set of data-based transformations to the social landscape that is both infrastructural and rhetorical in its form. This is a powerful combination of technological change allied with a cultural shift in which the possibilities of data become the basis for disruption and change. Kitchin presents us then with some difficult questions about how we comprehend such apparent change, how we might conceptualise it and, crucially, how these new conceptualisations of these developments might be historicised and contextualised. It has been noted by Patricia Owens (2015) that the relations between theory and history need to be rethought for the field of international relations to flourish. The same can certainly be said for work on data and metrics, here history and theory need to be used to give the analysis of the apparent big data revolution further context and conceptual depth. It is through such a focus that we might then 'open up to view the inherent politics and agendas of big data' (Kitchin 2014a: 127). Kitchin argues that this political agenda is important 'given that big data are reshaping how citizens and places are governed, organisations managed, economies work, and science is practised' (Kitchin 2014a: 127). I hope here to take on Kitchin's challenge and to open up this political agenda through a conceptual engagement with the relations between metrics and power. As this would suggest, we need to think carefully about the cultural politics that afford such an apparent data revolution, and it is to this part of Kitchin's challenge that I focus the remainder of this chapter.

Metrics *and* Power

To lend even further context and to respond to the challenges laid out above, the rise and intensification of metrics can also be thought of in terms of debates on governmentality and power. To place metrics within broader political structures and constellations, this opening chapter attempts to understand metrics within the context of what is often referred to as neoliberalism and neoliberalisation. Stuart Hall (2011: 711) explains that in general terms neoliberalism 'borrows and appropriates extensively from classical liberal ideas; but each is given a further "market" inflexion and conceptual revamp'. The result, for Hall (2011: 723), is that we become locked into a perpetual engagement with neoliberal thinking for, he claims, 'what neoliberalism wants to engineer is permanent revolution'. The notion of the market, Hall points out, is central to understanding the move to neoliberalism. Indeed, as this chapter will discuss, competition through markets is often seen to be central to the neoliberal 'art of government' (Foucault 2007). Neoliberalism, according to Jodi Dean (2009: 52), is the 'reformatting of social and political life in terms of its ideal of competition within markets'. Dean is not alone here, we can actually locate a long list of thinkers that draw a similar line. If we can move beyond the gendered expression they provide, Dardot and Laval (2013: 256, italics in the original) make the poignant and bold proclamation that 'neoliberal man is *competitive* man, wholly immersed in global competition'. They also claim that the 'principle characteristic of neo-liberal rationality is the generalization of competition as a behavioural norm and of the enterprise as a model of subjectivation' (Dardot and Laval 2013: 4). In fact, in slightly more nuanced terms, they also propose that neoliberalism 'can be defined as the set of discourses, practices and apparatuses that determine a new mode of government of human beings in accordance with the universal principle of competition' (Dardot and Laval 2013: 4). Here then competition is a 'behavioural norm' that permeates through social structures to become a central principle of the 'government' of people. Similarly, David Harvey (2005: 65) also proposes that for neoliberalism 'competition—between individuals, between firms, between territorial entities (cities, regions, nations, regional groupings)—is held to be a primary virtue'. Competition then is not just an organising principle but is also a virtue. A point which was echoed by Philip Mirowski (2013: 92) in his claim that 'competition is the primary virtue, and solidarity a sign of weakness'.

Clearly then, competition is seen, generally, to be a central focus of neoliberal thinking or of a neoliberal approach to governance. We can turn to a number of further accounts to cement such a conclusion. Neoliberalism, as Peck and Tickell (2002: 33) have observed, can be thought of as promoting an almost religious 'commitment to the extension of markets and logics of competitiveness'. Competition, played out in the form of markets, is not, it would seem, just a virtue but is based upon a type of religious faith. Yet Wendy Brown (2015b: 31; italics in the original) makes an important distinction: She points out that what we are seeing here is not necessarily simply the spreading of markets. Neoliberalism is not marketisation; instead, she suggests that 'neoliberal rationality disseminates the *model of the market* to all domains and activities'. This means that even where money may not be directly involved you can still have the model of the market in action. So, it is not necessarily the market as such that spreads out into new domains, but the model of the market (and, of course, we might also question the very notion of markets in the first place, see Davies 2013 and Mirowski 2013). We have a complex concept to work with here, in which, to give us a point of departure, competition is pursued through a model of the market.

Having such a central anchor point is not always seen to be enough. Recently, Rajesh Venugopal has poured scorn on the concept of neoliberalism and the ways in which it has been deployed. Venugopal (2015: 183) is concerned with the incoherence of the concept as 'it lives on as a problematic rhetorical device that bundles together a proliferation of eclectic and contradictory concepts'. Venugopal's (2015: 182) conclusion is that when it comes to neoliberalism all we have left of any use is its 'descriptive shell'. Terry Flew (2014: 51–53) also raises some wide-ranging concerns about the diffuse, vague, and ill-defined uses of neoliberalism, and the tendency to use it as a kind of 'all-purpose denunciatory category' or as a reductive account of the 'way things are'. Given all of this, it is perhaps of little surprise that Jamie Peck (2013: 133) has described neoliberalism as a 'rascal concept'. Despite the fact that we have the common themes of the expansion of markets into new domains and the dogged pursuit of competition as a key organising principle, we need to think further about the details of the concept of neoliberalism and its relations with metrics. Whether or not we accept that neoliberalism is the driving force for various social transformations we can instantly see how metrics might have a key role to play in affording such a set of impulses and developments. Put simply competition and markets require metrics, measurement is needed for the differentiations required by competition.

This chapter, and book, proceeds cautiously with neoliberalism as a political economic context in order to highlight potential connections between the phenomena I explore and the broader power structures of which they are a part. It does not though assume neoliberalism to be an empirical fact, rather it is used as a concept here to think about the role of metrics in social ordering processes and as the cultural context or way of thinking in which the infrastructures of measurement are deployed—despite any limitations it is a useful concept for helping us to think in such terms. Peck (2013: 153) points out that 'the process of neoliberaliza-tion must not be a substitute for explanation; it should be an occasion for explanation'. In other words, we cannot explain things away by sim-ply uttering the word neoliberalism, concepts are never an alternative to explanation. Rather we need to use it to open up and explore these issues. Peck's contention here is that the concept of neoliberalism defines 'a problem space and a zone of (possible) pertinence, and as such represents the beginning of a process of analysis'. It is a concept then that marks out a terrain for questioning, it does not provide the answer to those questions. The concept of neoliberalism might provide a starting point, but it certainly should not be a conclusion in its own right. In keeping with Peck's suggestion, this book does not fall back upon neoliberalism as an explanation but rather draws upon it as a resource that enables us to place metrics within broader social and political conditions. The concept of neoliberalism is used to place metrics into a problem space rather than to offer a reductive explanation for their presence and use. Neoliberalism resides in the shadows, presenting questions that enable us to join some of the dots that are connected in this book.

However we decide to proceed with our notions of neoliberlisation, competition, and markets we can be certain that metrics are implicit in such power structures. David Harvey, for instance, notes the pursuit of technological means that enable markets to emerge or be strengthened—this is to highlight the importance of infrastructures to neoliberalism. Neoliberalism, Harvey (2005: 3) claims:

> holds that the social good will be maximized by maximizing the reach and frequency of market transactions, and it seeks to bring all human action into the domain of the market. This requires technologies of information creation and capacities to accumulate, store transfer, analyse, and use massive databases to guide decisions in the global marketplace. Hence neoliberalism's intense interest in the pursuit of information technologies.

In short, neoliberalism is likely to be based upon the pursuit of the expansion of the contemporary data assemblage. Given the interest in competition, a neoliberal approach is likely to aim to ramify the presence of networked technologies in practices and processes on an individual, organisational, and state level. A neoliberal approach is likely to pursue an intensification of data and the attendant industry of analytics—and then to build these into decision-making, choice selection, visibility, and valuation (which we will return to throughout the book, but most directly in Chap. 4). We can see then the potential importance of the data assemblage and the infrastructures and systems of measurement to neoliberalism—*systems of measurement and data extraction might be seen as the means of neoliberlisation.*

The problem which continues though, as the above references illustrate and as Wendy Brown (2015b: 20) makes clear in her recent book, is that neoliberalism remains 'too loose a signifier…there is temporal and geographical variety in its discursive formulations, policy entailments, and material practices'. We are presented then with the problem of variability. It is, she continues, 'globally ubiquitous, yet, disunified and nonidentical', it 'takes diverse shapes and spawns diverse content and normative details' (Brown 2015b: 21). Far from being a fixed term then, neoliberalism can account for variation, it is a concept for comparative analysis. As we shall see though, this variety is part of how this form of governance operates as it is adopted and mutates in different contexts—embracing and understanding such variation is part of the analytical capacity of this concept. For Brown, this is how we should approach neoliberalism. Brown's (2015b: 21) point is that an 'alertness to neoliberalism's inconsistency and plasticity cautions against identifying its current iteration as its essential and global truth and against making the story…a teleological one'. Neoliberalism, Peck (2013: 133) has also been quick to tell us, is changeable and adapts.

We should remind ourselves again, as Hall (2011: 708) points out, neoliberalism 'has many variants' for it is 'not a single system'—with the history of the relations between liberalism and neoliberalism being complex (see e.g. Jessop 2002) and even the history of the emergence of the concept being drawn through different genealogical lineages (see Gane 2014a). As such, we are likely to see metrics, as the mechanisms of competition, evolving differently and changing shape in different contexts. Neoliberalisation processes are even said to have something of a 'zigzagging character' (Peck 2010: 28), leading to Peck's preference for 'processual definitions of neoliberal*ization*' rather than 'static and taxonomic renderings of neoliberal*ism*' (Peck 2010: 19, italics in the original). Neoliberalism, Peck (2010: 20) contends, 'has always come in varieties'. Metrics are likely then to also be changeable, mobile, and varied as they interface with these broader processes.

So, despite these drawbacks, neoliberalism remains a potentially useful concept in the context of this book, largely because it shows how we might link metrics directly to the political formations of the day and to the historical genealogy that has led to these connections (but for an alternative account of the global genealogy of neoliberalism in relation to colonialism, see Venn 2009). It is not necessary, Peck has argued, for us to jettison the concept merely because of its failings. He puts this position and the limitations of the concept itself into stark terms:

> Just because neoliberalism is not a universal and ubiquitous phenomenon: just because it is not monolithic, unilogical, and free of contradiction; just because it is not teleologically trained on a history-ending form of global

convergence; just because it is not an unstoppable, self-replicating system, utterly impervious to outside influences and immune from effective contestation…just because neoliberalism does not, indeed cannot, satisfy these absolutist, hyperbolic criteria, this does not mean that it is a figment of the (critical) imagination (Peck 2010: 15)

Despite these problems, for Peck the concept can still be used to engage in productive encounters with social and political issues and questions. It is a concept that still has analytical purchase. Let us not forget Peck's (2010: 15) observation that the 'tangled usage of neoliberalism may be telling us something about the tangled mess of neoliberalism itself'. We can proceed, perhaps tentatively, by taking Jamie Peck's (2010: 20) advice and thinking of neoliberalism as denoting a 'problem space'. We may then be able to understand metrics within this problem space and to see how they are implicit in the promises and visions of neoliberalism and in the actualities or tangled messes of the variegated unfolding of neoliberalisation processes. In short, the limitations of the concept of neoliberalism might help us in sharpening our analysis of the role of metrics. How do metrics limit the deployment of neoliberal dreams? How do metrics fulfil neoliberal visions? How are metrics evoked in the promises presented in the pursuit of neoliberal rationalities? What part do metrics play in the contradictions inherent in neoliberalisation? These are just some of the questions that we might ask.

One of the arguments I make in this book is that it is in the circulation of the outputs of systems of measurement that we might see the variability of the 'variegated' art of neoliberal governance (see Brenner et al. 2010; Gane 2012). It is in the selection, prioritisation and force of these circulations of metrics that we might find the operation of power or where we might locate what Nicholas Gane (2012) has crucially described as a form of 'surveillance' through 'competition'. Looking closely at systems of measurement and the use and circulation of metrics might well then reveal something about the variations and messiness in the broader political order.

We need to keep this adaptability and variability in mind if we are to try to place metrics within such a political economy. I have suggested that metrics are the means and mechanisms by which competition can develop and spread across different spheres of society. As such, we should be looking to explore the variability in these forms of measurement, to investigate variability across time and geographies, and to embrace rather than reject unevenness and inconsistency in our analysis—this unevenness might reveal something of the realities of the political projects of which metrics may be a part. Underpinning this position is the simple claim that in order to understand these far reaching questions about power and governance we need to also understand the systems of measurement that reside within them.

Competition and Metrics

As the above suggests, systems of measurement are crucial in the realisation and deployment of what might be thought of as neoliberal political formations and the processes of neoliberlisation. Put simply, measurement is needed for competition to exist—systems of measurement come hand in hand with what Peck and Tickell (2002: 48) refer to as the 'deliberate extension of competitive logics'. Measurement is needed to enable competitors to be judged and for hierarchies of winners and losers to be created. Systems of measurement provide the mechanisms by which that competition can be enacted.

Given neoliberalism's central ethos of competition (Davies 2014), measurement can be seen to be a crucial part of the social fabric. Foucault (2008: 147), standing at his tape recorder- laden lectern back in 1979, argued that:

> [the] society regulated by reference to the market that the neo-liberals are thinking about is a society in which the regulatory principle should not be so much the exchange of commodities as the mechanisms of competition. It is these mechanisms that should have the greatest possible surface and depth and should also occupy the greatest possible volume in society.

This is the shift from markets based upon exchange to those based upon competition (see also Brown 2015b: 36). In short, neoliberalism requires and pursues forms of marketised competition—competition, of course, requires measurement. Foucault's secondary point here is that these mechanisms need to the 'greatest possible surface', and as such the basis for competition needs to reach as many objects, people, and organisations as possible. This is about increasing the reach of governance by making more things measurable. This expansion is fuelled by what Dean (2009: 55) describes as the 'fantasy of free trade' and the 'fantasy promises that an unfettered market benefits everyone'. The promise and dreams of the potential of marketised competition then become the rationale and driver for neoliberal governance, and therefore for the spread of metrics as the mechanisms for that competition.

The observation I'd like to reiterate at this juncture though is that metrics are needed for competition to function. Metrics are the very mechanisms and apparatus by which competition can be realised; metrics afford differentiations to be created and inequalities to be cemented. And beyond this, the expansion of systems of measurement is needed for competition to spread out into the social world. It is in the very relations between metrics and competition that we find these apparently 'cold intimacies' of contemporary capitalism (Illouz 2007) and where neoliberalism is made and experienced. Neoliberalism has been seen to adapt and to remake itself in light of its conditions (Peck and Tickell 2002: 53; Peck 2013: 144). Metrics are likely to be at the forefront of these redesigns.

Will Davies argues that 'the common thread in all of this—and what makes the term "neoliberalism" a necessary one—is an attempt to replace political judgement with economic evaluation, including, but not exclusively, the evaluations offered by markets' (Davies 2014: 3). Davies' claim is that a key feature of neoliberalism is an attempt to move towards economic evaluation, such evaluation requires some form of measure, particularly in the instances where the shift is towards market-based evaluations in which comparison is central. Davies' point is that there is a general rise in 'hostility' towards 'ambiguity' which is accompanied by a 'commitment to the explicitness and transparency of quantitative, economic indicators…Neoliberalism is the *pursuit of the disenchantment of politics by economics*' (Davies 2014: 4, italics in the original). Clearly then, in order for neoliberalism to function, for its hostility towards ambiguity to be placated, the means by which indicators can be located are a necessity. Indeed, it has been argued that the establishment of indicators is crucial part of contemporary governance (Rottenburg and Merry 2015) and of the establishment of the very practice of measurement itself (Desrosières 2015). Here, quantitative properties are foregrounded as indicators of value and worth under the pursuit of economic judgments (see Chap. 4). We might see this then as being the moment for metrics to flourish as a key component of the functioning and ordering of the social world (see Chap. 2). Neoliberal governance demands indicators and the means by which those indicators can be compared and contrasted, it is the implementation of systems of measurement and the pursuit of measures that makes this possible. Similarly, if we think of this more as an ongoing process of 'variegated neoliberlization' (Brenner et al. 2010) then it is the expansion of systems of measurement that enable these ongoing processes to unfold and escalate.

The Variegated Mechanisms of Competition

It is here that we need to pause to revisit the variable types of terrain that this concept of neoliberalism is being applied within. As we have already seen, neoliberalism, in the finer-tuned and richer accounts, is not a monolithic and coherent thing, far from it in fact. In David Harvey's (2005: 19) history of neoliberalism, for instance, he identifies a 'creative tension between the power of neoliberal ideas and the actual practices of neoliberalization'. The acknowledgement of such a distinction has led Peck and Tickell (2002) to write of the 'mutations' of neoliberalism. We see here then the move from neoliberalism as a concept to neoliberalism as a set of policy decisions and governmental manoeuvres. As a result, perhaps unsurprisingly, they argue that the only way to proceed with an analysis of neoliberal formations is to think in terms of processes and to be 'attentive to *both* the local peculiarities *and* the generic features of neoliberalism' (Peck and Tickell 2002: 41, italics in the original). It is in finding such a balance between the shared features of neoliberalism and the unusual specifics, Peck and Tickell argue, that we can develop a more nuanced understanding of its application and analytical value.

We might then think, in the context of this book, that it is important to think of variations in how people are measured and the differences in what those metrics are used for. It is for this reason that it has been argued that it is important to explore the 'contextual embeddedness of neoliberal restructuring projects' (Brenner and Theodore 2002: 4). As Peck and Tickell further explain, and echoing the types of analytical problems that have been discussed at length (famously by Mills 1959), this 'means walking a line…between producing, on the one hand, overgeneralized accounts of a monolithic and omnipresent neoliberalism, which tend to be insufficiently sensitive to its local variability and complex internal constitution, and on the other hand, excessively concrete and contingent analyses of (local) neoliberal strategies, which are inadequately attentive to…substantial connections' (Peck and Tickell 2002: 34). This leaves us aiming to see neoliberalism as a concept that needs to be seen in both generic and specific terms, for it has both overarching tendencies and an uneven geography of development (see Harvey 2005: 87–119; Peck 2010). This is a difficult line to walk as we are drawn both to the common and the unusual, the detail and the general, the specific and the non-specific, the macro, micro, and the meso—we are working across very different scales and with a concept that does very different things. For the moment, it is perhaps sufficient to acknowledge, at the least, that neoliberalism is not being used as a catch-all term here. Rather, there is a tension at the centre of the concept as its broad principles come to be applied in different ways in different places and at different times. As we have seen, one way to think of neoliberalism is as a concept in or of tensions—the generic and the local is one, but we can also add in the tensions around intervention and the promotion of competition amongst others (for an outline of the tensions and contradictions of neoliberalism, see Harvey 2005: 67–70, 79–81).

Peck's position would again suggest to us that we should be looking at the underpinning variability in the systems of measurement that afford such unfolding processes. Indeed, it is worth restating Peck's point that neoliberalism doesn't exist in any pure form. For Peck (2010: 31), the 'actually existing worlds of neoliberalism are not pristine spaces of market rationality and constitutional order; they are institutionally cluttered places marked by experimental-but-flawed systems of governance, cumulative problems of social fallout, and serial market failure'. As this makes clear, according to Peck, neoliberalism is reinvented and reshaped in response to failures, obstacles, contestation, and mess. It is not a polished oven-ready mode of governance that is implemented in a social vacuum. It is the very, for Peck (2010: 24), non-achievability of the pure neoliberal project that is an important feature of the process and which leads to it being remade and reconstructed in different contexts. The neoliberal project, for Peck (2010: 16), is 'paradoxically defined by the very unattainability of its fundamental goal—frictionless market rule'; rather, it is the 'oscillations and vacillations around frustrated attempts to reach it that shape the revealed form of neoliberalism as a contradictory mode of market governance'. We might begin to wonder what part the generation and availability of metrics might play in such blockages and frustrations—we might be led to wonder how the neoliberal project is reshaped to suit the reach of the available infrastructures of measurement. This is where the imagined vision of neoliberalism becomes part of the processes of neoliberalisation. We see again the role of the promises of neoliberalism as a key driver to neoliberalisation—even though they are never likely to be achieved. Similarly, we might reflect on how the imagined power of metrics might be part of how those metrics are deployed. Peck (2010: 24) explains that:

> Roll-out neoliberalization, then, represents more than an attempt to remake the world in the image of markets, most actually existing alternatives having been weakened. Instead, it represents a series of far-from-perfect attempts to wrestle with the challenges and contradictions of governance in a malmarketized world.

Peck's point is clear; neoliberalisation might be the pursuit of certain market values, but what is interesting is not these 'market-utopian ideals' but how they play out as they are remade in response to the pressures and forces to which they are exposed. This is where we again might look at metrics as being a part of both this imperfect roll-out and the guiding promises of neoliberalisation. Similarly, Wendy Brown (2015b: 9) adds that, 'more than merely saturating the meaning or content of democracy with market values, neoliberalism assaults the principles, practices, cultures, subjects and institutions of democracy understood as rule by the people'. This is something more embedded than just markets, it is the raft of changes that comes with those market values. Brown's point is that the expansion of markets and market values goes well beyond any simple monetisation process. She argues that 'all conduct is economic conduct; all spheres of existence are framed and measured by economic terms and metrics, even when those spheres are not directly monetized' (Brown 2015b: 10). So to spread markets is not simply to monetise everything directly, but to frame everything in economic terms and to use metrics to pursue these interests. This suggests that *metric power* will vary dependent upon its roots in different infrastructures, ideological dreams, policy processes, promises, and political cultures.

What Should Be Measured and How? Metrics and Judgement

We can extend these thoughts still further. For Davies, we can find some space in which to progress with our analysis by thinking about the disenchanting role of economics and statistics. These he places as central to the establishment of neoliberal thinking. As he puts it:

> The positivist social sciences, along with various forms of 'governmentality' and statistics, seek to replace critique with technique, judgment with measurement, but they are constantly parasitical on higher order claims about *what* ought to be measured, and *how* it is legitimate to represent this objectively. (Davies 2014: 16, italics in the original)

The general shift then might be seen to be towards measurement as a replacement or substitute for more qualitative judgment, but this also requires shifts in expectations about what should be measured and how these measures might be presented as objective and trustworthy. As Porter (1995: ix) has put it, it is often understood that a 'reliance on numbers and quantitative manipulation minimizes the need for intimate knowledge and personal trust' (as I discuss in Chap. 2). In these circumstances, 'mechanical objectivity' provides 'an alternative to trust of personal knowledge' (Espeland 1997: 1108). The move towards technique and measure is ushered in not simply by systems of measurement but also by the framing of the metrics they produce as being legitimate, reliable, and fair. In other words, it is not just about what can technically be measured, but also the power of the claims about what is the right thing to measure and what that measure can reasonably be used for.

When thinking about these questions of what can be measured and how, we need to return again to notions of competition and to the spread of competitiveness. It is the pursuit of competition that often defines what is measured. Measurement and competition run hand in hand, in terms of having the capacity to justify one another. The question then becomes one of intervention. The question then is what role metrics play in interventions into competition or markets. With a neoliberal approach, interventions by the state or those in power only come to ensure that the conditions of competition can be established and maintained (for a critical overview of the problems of the separation of neoliberalism from liberalism, social change since 1979, and the absence of the citizen, see Brown 2015b: 47–78). That is to say that intervention is limited to 'safeguarding the conditions in which profitable competition can be pursued' (Hall 2011: 707). Safeguarding and defining the limits of measurement is an inevitable by-product of this. Under these conditions, Dean (2009: 51) suggests, the 'primary role of the state is to provide an institutional framework for markets' and 'creating markets in domains where they may not have previously existed'. Similarly, Wendy Brown (2015b: 28) also points towards this set of relations with 'economic policies' being forged with the 'root principle of affirming free markets'. As such, this is not a kind of governance based upon some sense of natural competition; rather, it is about fostering the conditions by which competition can be spread and maintained throughout the social world. Similarly, for Gordon (1991: 41), the market is not a 'natural social reality'; so it becomes, from this position, necessary for government to apply policies that enable markets to 'exist and function'. As Davies (2014: 29) further explains, 'once we are speaking of these deliberately constructed competitions, and not some existential or biological idea of emergent competition, we get a clearer view of the strange forms of authority which neoliberalism has generated and depended upon'. In such a set-up, with competition at its centre, 'metrics' and 'norms' are used to 'rank and distinguish worth' (Davies 2014: 35). This is about the calculation of worth;

it is about using measures to facilitate competition which then dictates what is seen to be valuable, successful, and desirable (I will discuss this in Chap. 4).

Clearly then, the rules of the competition are powerful in defining worth. The metrics used, how they are compared, the outcomes produced, the differences that are made visible, all become powerful in shaping social outcomes, chances, and perceptions. If judgement is bypassed by technique, then the metric becomes powerful in ordering the social world and in decision-making. It is perhaps no surprise that this has not suddenly appeared from nowhere but that 'rhetorics and theories of *competition and competitiveness* have been central to neoliberal critique and technical evaluations from the 1930s onwards' (Davies 2014: 37, italics in the original). The way that competition is theorised, the way it is set up, will have significant consequences for the outcomes and their reverberations. Metrics then play a central role in the formations of neoliberalism and its limits. Systems of measurement could be seen here to define competition, to decide what can measured and how, and thus to shape social outcomes where they are based on calculation and market-based competition. Systems of measurement are the means by which the shift can be made towards calculation and away from judgment and critique. But, it is important that we see these metrics as cultural and political objects as well as being technical and infrastructural by-products. They always have a purpose—or they are given one after the fact. They are never neutral. So we should also see the reach of metrics as being about the cultures and ways of thinking that justify and seek to expand those measures.

Foucault now famously described neoliberalism in terms of the 'the inversion of the relationships of the social to the economic' (Foucault 2008: 240; see also Gane 2012). At the end of his lectures on *Security, Territory, Population*, Foucault (2007: 346) proposes that 'competition will be allowed to operate between private individuals, and it is precisely this game of the interest of competing private individuals who each seek maximum advantage for themselves that will allow the state, or the group, or the whole population to pocket the profits, as it were'. Foucault is not speaking specifically about neoliberalism in this passage, although it foreshadows the following year's lecture series on what Foucault (2008) calls the neoliberal 'art of government' (see also Gordon 1991: 14); yet the passage gives a sense of the direction of flight that Foucault observes. We can see that, for Foucault, historically there is a shift towards competition as an ordering principle in the social world, and that calculation and measurement are central to the facilitation of these formations (an overview of the issue of calculation in the 1978 lectures is provided by Elden 2007). The type of competition that is being highlighted here is highly individualising. As Foucault's accounts would suggest, this is about private individuals competing with one another. Although as Davies (2015b) has very recently identified, simplified notions of individualisation have been problematised by a resurgence of 'the social', in the form of 'social enterprise' and 'social media', that brings with it new opportunities for exercising power.

Despite this apparent resurgence in notions of the social, under neoliberal conditions the individual comes to seek advantages, to get an edge, to compete and to win. Thus the neoliberal subject is likely to be metric focussed. This has significant consequences, particularly in terms of the apparent pursuit of increasing inequality and the insecurity that this fosters (see Lazzarato 2009)—this is what Brown (2015b: 28) refers to as 'intensified inequality'. The power shifts into the hands of those who are able to use numbers to facilitate competition and thus to cement social inequalities. The role of experts in enabling competition and the knowledge that is required to let it function or to be successful in these competitions is crucial. As Davies (2014: 30) explains:

> In a world organized around the pursuit of inequality, that is, by an ethic of competitiveness, these experts are able to represent the world in numerical hierarchies of relative worth. It is not just cold instrumental reason that underpins the authority of economics in the neoliberal state, but also its capacity to quantify, distinguish, measure and rank, so as to construct and help navigate a world of constant, overlapping competitions

The pursuit of inequality is a key property of neoliberal thinking. The effects of neoliberalism are certainly not felt equally [see e.g. the discussion of 'racial neoliberalism' in Kapoor (2013) or the discussion of gender and neoliberalism in Scharff (2014) or Oksala (2013)]. Competition inevitably creates unequal outcomes. Competitions are designed for that very purpose. But there is no single logic to these competitions, rather they overlap in our lives, acting upon us from different angles, measuring our value in lots of different ways. Endlessly ranking us. Brown (2015b: 38; see also page 41) similarly identifies how such an interest in 'human capital features winners and losers, not equal treatment or equal protection'. The emphasis is upon the experts who are able to find the measures that are deemed to count and to then use these and interpret them for an audience of interested parties. As Davies points out here, neoliberalism is made possible by the ability to rank. 'The pragmatic utility of economic methodologies', Davies continues, 'is to provide common *measures* and *tests* against which differences in value can be established' (Davies 2014: 30, italics in the original). Measures facilitate differentiation, especially on the grounds of value. Measures enable the production of winners and losers. As Brown (2015b: 30) puts it, neoliberalism is an 'order of normative reason that, when it becomes ascendant, takes shape as a governing rationality extending a specific formulation of economic values, practices, and metrics to every dimension of human life'. It is through metrics, and expansive systems of measurement and analytic expertise, that various types of competition are made real. Put in these terms, it is clear why understanding systems of measurement is so pressing, not least because metrics are central to facilitating the forms of competition that define neoliberal formations. For Foucault, 'the art of government is deployed in a field of relations of forces' (Foucault 2007: 312) in which the 'calculation of forces' (Foucault 2007: 295) is central. Understanding how those 'relations of forces' are calculated and then how they come to be felt acting upon individuals is the step that still needs to be taken.

Verifications and Truth-making

What we have so far then is both a way of thinking and a means by which that way of thinking can become *the way of the world*. Foucault (2007: 286) unpicks this 'art of government' in some detail, to show 'how this appearance of governmental reason gave rise to a certain way of thinking, reasoning and calculating'. In this sense, there was a particular form of governmental reason that developed into a certain way of thinking with regard to calculation. Thus, Foucault places such reasoning into a long genealogy that stretches back beyond the establishment of more recently advancing means of measuring people (see Chap. 2). Foucault tracks this back to the start of the seventeenth century where, he says, 'we see the appearance of a completely different description of the knowledge required by someone who governs' (Foucault 2007: 273). This type of emergent governance, he claimed, required the development of an 'apparatus of knowledge' (Foucault 2007: 275). This was a form of governance based upon knowledge or information about those being governed. As Foucault put it, 'knowledge of the things that comprise the very reality of the state is precisely what at the time was called "statistics"' (Foucault 2007: 274). We will discuss the history of statistics in Chap. 2, but for the moment it is worth noting that the type of use of metrics about which we are talking here can be seen to have its roots much earlier than we might naturally assume—particularly when we are seduced by the glossy and provocative rhetoric of big data. This type of art of government was based upon 'strategies and tactics' in the 'analysis of the structures of power' (Foucault 2007: 216), and is something of an unfolding set of concerns rather than a sudden epoch. This was seen in the long formation of an 'art of government' based upon the 'government's consciousness of itself' (Foucault 2008: 2) and the creation of infrastructures of knowledge about people. Analytical and strategic in its focus and in its understanding of its own practices, this is the vision of governance and power that Foucault was trying to prise open in his lectures.

As I have mentioned, for Foucault the move is towards a form of knowledge-based government in which intervention is merely based around the facilitation of competition. As such, interventions are often based upon what the measures reveal. For Foucault, such an approach creates limits and then shapes what is seen to be possible or truthful. For Foucault (2008: 17), 'the possibility of limitation and the question of truth are both introduced into governmental reason through political economy'. This then is a way of thinking that is based upon analytics, and upon the use of knowledge to set up limitations and to shape notions of truth—what he later calls 'regimes of truth' (Foucault 2014: 93). What Foucault means by this becomes clearer a little later in his 1979 lecture series when he talks of the market as being a 'site of veridiction' (Foucault 2008: 32–33). Here markets are seen to legitimise and authorise. Thus, markets verify truths and set limitations and possibilities. If we are thinking ahead to some of the arguments I develop in this book, the metrics used to facilitate competition come to reinforce themselves and to justify their own presence. Stephen Stigler (1986: 3) draws a similar conclusion in his history of statistics with his claim that 'many measurements carry with them an implicitly understood assessment of their own accuracy'. For Foucault though, this is not just about measures; it is about how those calculations are manifest in markets. Measures, in the form of markets, verify and legitimate themselves. Measures define what is true and then are used to verify that truth. This can take various forms, but the pursuit here is of a 'strategic logic' (Foucault 2008: 42) that adheres to the rationalities of the competition and its outcomes. An approach, then, that plays by and anticipates the rules of the competition. This is an approach to the social world that values 'reflection, analysis and calculation' (Foucault 2008: 59). In short, Foucault tracks the emergence of a form of governmentality that analyses itself and others (see also Rose 1999: xxi–xxiii).

Given this focus upon strategic logic, it is not surprising that for Foucault (2008: 118) competition is again seen to be the crucial component of neoliberal governmentality. This time we can add though that these competitions bring their own implicit logic. As he describes it:

> Competition is an essence...Competition is a principle of formalization. Competition has an internal logic; it has its own structure. Its effects are only produced if this logic is respected. It is, as it were, a formal game between inequalities; it is not a natural game between individuals and behaviours (Foucault 2008: 120)

Competition, according to Foucault, brings with it its own logic and structure. Foucault describes this as a formal game that works on inequalities. Marketised competition creates self-reinforcing truths and limits social life through its rules and outcomes. Again, the emphasis here is upon the engineering of competition rather than it being some natural order that is adhered to. Yet, competition only works if its logic is respected—to resist, to genuinely resist, is to defy this logic. Because this is formalised rather than natural competition, so the rules of the game need to be written and established. Foucault describes this as the implementation of 'competitive mechanisms' which come to 'play a regulatory role' (Foucault 2008: 145). Intervention, as we have seen, then comes in the form of the production and maintenance of competitive mechanisms and to ensure those rules are followed. We should perhaps conclude that such competitive mechanisms are based around the systems of measurement that facilitate ranking and ordering. Metrics then can be seen as the mechanisms of competition, they are the components that give competition its coherence and enable comparability. Metrics reinforce the logic and outcomes of competition, they can be used to verify truths and to cement limits. Interventions are likely to be interventions that reinforce those metrics and maintain the coherence and viability of the rankings they produce.

I do not intend to use this introduction, or indeed this book, to elaborate a full history or conceptual account of neoliberalism. As I have suggested so far, neoliberalism is not a straightforward concept with a clear lineage or definition. But we have seen here that competition, in various forms and with varying effects, is a key feature of these visions of neoliberalism. As such, we should encounter the rise of metrics in this context by connecting them with the broader political desire to measure, differentiate, and rank. There are no doubt connections here that need to be mined. To understand neoliberalism and neoliberlisation will be to understand the systems of measurement that act as the mechanisms of competition. Such accounts enable us to further cement the context in which metric power is realised. For the purpose of this book, I set this as the background and use it to suggest that metrics are powerful in the current conditions of which they are a part, and therefore metric power is a pressing concern for anyone who is interested in contemporary social and cultural constellations of power. Here the concept of neoliberalism is used to suggest just how deeply metrics are implicated in wider power structures and in the very ordering of the social world and individual lives—it enables us, if used sensitively and critically, to think through the wider role of metrics in the contemporary setting on different scales and in different geographical contexts. As I conclude in Chap. 5, borrowing from Foucault's (2007: 45) analysis, metric power is centrifugal in form. It spreads outwards through the social world. Part of the reason for this, I would suggest, is that it rides upon the back of processes of neoliberlisation which are defined by the pursuit and spread of competition across the various spheres of social life. To spread competition outwards is first to spread the reach of metrics. Stuart Hall (2011: 728), despite some hesitancies and caveats, goes as far as to suggest that neoliberalism could be seen to be hegemonic in its scope. He is not alone. David Harvey and Philip Mirowski have both commented on this sinking of neoliberal thinking and attitudes into the background of everyday life in the form of common sense and hard to dispute notions (see Harvey 2005: 3; Mirowski 2013: 89–156). This makes the understanding of the role of metrics in power even more pressing; we need to understand what I will call metric power with some urgency, particularly as it continues to soak into the fabric of the social world.

The Rest of the Book

Lisa Adkins and Celia Lury's (2012: 15) notable edited volume on the relations between value and measure is something I will return to, but at this opening moment it is interesting to note their point that, despite its scale and importance, 'social scientists are only just beginning to engage with this emergent economy of data including the politics of measurement attached to it'. Despite a number of telling interventions, which I discuss in Chap. 2 and throughout this book, this still needs our attention. It is to this project that this book is dedicated, not in pursuit of a resolution but in doing the conceptual groundwork that might allow such a broader project to flourish. The concept of metric power that I have begun to outline in this introduction is the means by which these aims can be achieved. The rest of the book is dedicated to its elaboration and to fleshing out its analytic potential.

Metrics provide the mechanisms for the realisation of competition and the means by which that competition can spread across the social world. It has been said, as we have seen, that the aim of neoliberalism is 'to shape subjects to make them entrepreneurs capable of seizing opportunities for profit and ready to engage in the constant process of competition'

(Dardot and Laval 2013: 103). We need not necessarily fully buy into the concept of neoliberalism to see that competition and judgement through metrics is now rife (see Chap. 2). Indeed, it is hard to contest the notion that metrics have assumed a significant role and power in contemporary social life. With developments such as big data, which are both technical and rhetorical in their form, we have what might be seen as the perfect conditions for the expansion of competition—and thus potentially, for increased neoliberalisation. Despite any disagreements we might have about the conceptual terminology, we might at least agree that there is plenty to suggest that society is continuing with its long held direction towards metricisation (see Chap. 2).

The central argument of this book is that understanding the intensification of measurement, the circulation of those measures and then how those circulations define what is seen to be possible, represents the most pressing challenge facing social theory and social research today. A focus on the relations between measurement, circulation, and possibility is one way of illuminating and revealing the intricate linkages between metrics and power. The concept of metric power is built upon the exchanges between these three movements. Such an approach enables us to ask how these circulating measures define and prefigure what is possible and what is imagined to be possible.

In order to face this challenge and to open up such a set of conceptual possibilities, this book will take each of these three components—measurement, circulation, possibility—in turn. Each will be developed in isolation, to draw out the key issues, before they are seen in combination in the concluding chapter. The book as a whole is structured a little like a piece of music, with three movements and a coda. This coda, Chap. 6, adds an extra bodily dimension to metric power and is used to explore how the power of metrics resides within their affective capacities.

Moving through these three spheres in this way—from measurement to circulation to possibility—will hopefully show how they can be seen as discrete elements whilst also then illustrating how they connect together. The format of the book suggests this is a linear process, with measurements being generated before circulating out into the world and creating possibilities. To a certain extent, there is a logic to this linear flow, and it is why I have placed the chapters in this order. However, I'd like to be clear from the outset in stating that the relations between these three elements are unlikely to be linear but are far more likely to be based upon recursive and recombinant imbrications and connections—for instance, what is seen to be possible may feedback into what is measured and so on. My hope is that this will become clearer as the book progresses, particularly as the discussion moves to questions of circulation in Chap. 3. But as a starting point, the connections and relations that make up and afford *metric power* are not a clear set of linear processes with clear outcomes; rather, they are embedded into all sorts of infrastructural, organisational, corporeal, and governmental feedback loops. Let us begin to join the dots by focussing upon measurement.

References

Beer, D. (2013). *Popular culture and new media: The politics of circulation*. Basingstoke: Palgrave Macmillan.

Beer, D. (2015a). The apple watch and the problem of our creeping connectivity. *Sociological Imagination*. Accessed July 6, 2015, from http://sociologicalimagination.org/archives/17447

Beer, D. (2015b). Productive measures: Culture and measurement in the context of everyday neoliberalism. *Big Data and Society, 2*(1), 1–12.

boyd, d., & Crawford, K. (2012). Critical questions for big data: Provocations for a cultural, technological and scholarly phenomenon. *Information Communication and Society, 15*(5), 662–679.

Brenner, N., Peck, J., & Theodore, N. (2010). Variegated neoliberalization: Geographies, modalities, pathways. *Global Networks, 10*(2), 182–222.

Brenner, N., & Theodore, N. (2002). Cities and the geographies of "actually existing neoliberalism". In N. Brenner & N. Theodore (Eds.), *Spaces of neoliberalism: Urban restructuring in North America and Western Europe* (pp. 2–32). Oxford: Blackwell.

Brown, W. (2015b). *Undoing the demos: Neoliberalism's stealth revolution*. New York: Zone Books.

Dardot, P., & Laval, C. (2013). *The new way of the world: On neoliberal society*. London: Verso.

Davies, W. (2013). When is a market not a market? 'Exemption', 'externality' and 'exception' in the case of European State Aid rules. *Theory Culture and Society, 30*(2), 32–59.

Davies, W. (2014). *The limits of neoliberalism*. London: Sage.

Davies, W. (2015a). *The happiness industry: How the government and big business sold us well-being*. London: Verso.

Davies, W. (2015b). The return of social government: From 'socialist calculation' to 'social analytics. *European Journal of Social Theory*. Online first. doi: 10.1177/1368431015578044.

Day, R. E. (2014). *Indexing it all: The subject in the age of documentation, information, and data*. Cambridge, MA: MIT Press.

Dean, J. (2009). *Democracy and other neoliberal fantasies: Communicative capitalism and left politics*. Durham, NC: Duke University Press.

Deloitte. (2015). *Global human capital trends 2015: Leading in the new world of work*. London: Deloitte University Press.

Desrosières, A. (2015). Retroaction: How indicators feed back onto quantified actors. In R. Rottenburg, S. E. Merry, S. J. Park, & J. Mugler (Eds.), *The world of indicators: The making of governmental knowledge through quantification* (pp. 329–353). Cambridge: Cambridge University Press.

Elden, S. (2007). Governmentality, calculation, territory. *Environment and Planning D: Society and Space, 25*(3), 562–580.

Espeland, W. N. (1997). Authority by the numbers: Porter on quantification, discretion, and the legitimation of expertise. *Law and Social Inquiry, 22*(4), 1107–1133.

Espeland, W. N., & Sauder, M. (2007). Rankings and reactivity: How public measures recreate social worlds. *American Journal of Sociology, 113*(1), 1–40.

Espeland, W. N., & Stevens, M. L. (2008). A sociology of quantification. *European Journal of Sociology, 49*(3), 401–436.

Flew, T. (2014). Six theories of neoliberalism. *Thesis Eleven, 122*(1), 49–71.

Foucault, M. (2002b). *Power: Essential works of Foucault 1954–1984,* (Vol. 3). London: Penguin.

Foucault, M. (2007). *Security, territory, population: Lectures at the Collège de France 1977–1978*. Basingstoke: Palgrave Macmillan.

Foucault, M. (2008). *The birth of biopolitics: Lectures at the Collège de France 1978–1979*. Basingstoke: Palgrave Macmillan.

Foucault, M. (2013). *Lectures on the will to know: Lectures at the Collège de France 1970–1971 and Oedipal Knowledge*. Basingstoke: Palgrave Macmillan.

Foucault, M. (2014). *On the government of the living: Lectures at the Collège de France 1979–1980*. Basingstoke: Palgrave Macmillan.

Gane, N. (2012). The governmentalities of neoliberalism: Panopticism, post-panopticism and beyond. *Sociological Review, 60*(4), 611–634.

Gane, N. (2014a). The emergence of neoliberalism: Thinking through and beyond Michel Foucault's lectures on biopolitics. *Theory Culture and Society, 31*(4), 3–27.

Gordon, C. (1991). Governmental rationality: An introduction. In G. Burchill, C. Gordon, & P. Miller (Eds.), *The Foucault effect* (pp. 1–51). Chicago: The University of Chicago Press.

Hall, S. (2011). The neo-liberal revolution. *Cultural Studies, 25*(6), 705–728.

Harvey, D. (2005). *A brief history of neoliberalism*. Oxford: Oxford University Press.

Illouz, E. (2007). *Cold intimacies: The making of emotional capitalism*. Cambridge: Polity Press.

Jessop, B. (2002). Liberalism, neoliberalism, and urban governance: A state-theoretical perspective. In N. Brenner & N. Theodore (Eds.), *Spaces of neoliberalism: Urban restructuring in North America and Western Europe* (pp. 105–125). Oxford: Blackwell.

Kantor, J., & Streitfeld, D. (2015, August 15). Inside Amazon: Wrestling big ideas in a bruising workplace. *The New York Times*. Accessed August 19, 2015, from http://www.nytimes.com/2015/08/16/technology/inside-amazon-wrestling-big-ideas-in-a-bruising-workplace.html?_r=0

Kapoor, N. (2013). The advancement of racial neoliberalism in Britain. *Ethnic and Racial Studies, 36*(6), 1028–1046.

Kitchin, R. (2014a). *The data revolution: Big data, open data, data infrastructures & their consequences*. London: Sage.

Kittler, F. (2006). Number and numeral. *Theory Culture and Society, 23*(7–8), 51–61.

Lazzarato, M. (2009). Neoliberalism in action: Inequality, insecurity and the reconstitution of the social. *Theory Culture and Society, 26*(6), 109–133.

Mills, C. W. (1959). *The sociological imagination*. Oxford: Oxford University Press.

Mirowski, P. (2013). *Never let a serious crisis go to waste: How neoliberalism survived the financial meltdown*. London: Verso.

Oksala, J. (2013). Feminism and neoliberal governmentality. *Foucault Studies, 16*(1), 32–53.

Owens, P. (2015, January 31). Interview—patricia owens. *E-International Relations*. Accessed February 10, 2015, from http://www.e-ir.info/2015/01/31/interview-patricia-owens/

Peck, J. (2010). *Constructions of neoliberal reason*. Oxford: Oxford University Press.

Peck, J. (2013). Explaining (with) neoliberalism. *Territory, Politics, Governance, 1*(2), 132–157.

Peck, J., & Tickell, A. (2002). Neoliberalizing space. In N. Brenner & N. Theodore (Eds.), *Spaces of neoliberalism: Urban restructuring in North America and Western Europe* (pp. 33–57). Oxford: Blackwell.

Porter, T. M. (1995). *Trust in numbers: The pursuit of objectivity in science and public life.* Princeton, NJ: Princeton University Press.

Rose, N. (1991). Governing by numbers: Figuring out democracy. *Accounting Organization and Society, 16*(7), 673–692.

Rose, N. (1999). *Governing the soul: The shaping of the private self* (2 ed.). London: Free Association Books.

Rottenburg, R., & Merry, S. E. (2015). A world of indicators: The making of governmental knowledge through quantification. In R. Rottenburg, S. E. Merry, S. J. Park, & J. Mugler (Eds.), *The world of indicators: The making of governmental knowledge through quantification* (pp. 1–33). Cambridge: Cambridge University Press.

Scharff, C. (2014). Gender and neoliberalism: Exploring the exclusions and contours of neoliberal subjectivities. *Theory, Culture and Society.* Accessed August 20, 2015, from http://theoryculturesociety.org/christina-scharff-on-gender-and-neoliberalism/

Schiller, D. (2014). *Digital depression: Information technology and economic crisis.* Urbana, IL: University of Illinois Press.

Stables, J. (2015, April 9). US Insurance firm offers trackable Fitbit for lower insurance premiums. *Wearable.* Accessed January 12, 2015, from http://www.wareable.com/wearable-tech/us-insurance-firm-offers-trackable-fitbits-for-lower-premiums-1032

Stigler, S. M. (1986). *The history of statistics: The measurement of uncertainty before 1900.* Cambridge, MA: Belknap Press of Harvard University Press.

Stiglitz, J., Sen, A., & Fitoussi, J. P. (2010). *Mismeasuring our lives: Why GDP doesn't add up.* New York: The New Press.

Thrift, N. (2005). *Knowing capitalism.* London: Sage.

Venn, C. (2009). Neoliberal political economy, biopolitics and colonialism: A transcolonial genealogy of inequality. *Theory Culture and Society, 26*(6), 206–233.

Venugopal, R. (2015). Neoliberalism as concept. *Economy and Society, 44*(2), 165–187.

2

Measurement

How much are the oceans worth? This might seem an odd or perhaps even pointless question, but it does not stop those who wish to place a number on everything. The answer—$24 trillion (WWF 2015). Not only is the question an incomprehensible abstraction, but so is the answer. An incongruous question is answered with a monetary value that can't even be imagined. Although we should note that in this instance it is a figure that is being used with the aim of encouraging the protection of marine environments—it would seem that valuing the oceans is then a strategy to speak to those who wish to understand through economic valuations. Of course, this has been challenged. Charles Eisenstein (2015) has written a beautifully succinct piece that cuts directly to the problems of measuring the monetary value of oceans. Eisenstein does a far better job of problematising this valuation than I would be able to, but this example is useful in illustrating the scale of measurement today and the depth of the impulse or desire to measure.

In his book on the politics surrounding large numbers, Alain Desrosières suggests that there are broadly two types of questions or controversies that are raised in relation to the use of statistical measurement as providing objective and indisputable numerical accounts of the social world. For Desrosières, the questions that are raised depend on whether

© The Editor(s) (if applicable) and The Author(s) 2016
D. Beer, *Metric Power*, DOI 10.1057/978-1-137-55649-3_2

the concern is with the measurement itself or with the object that is being measured. As he explains:

> In the first case, the reality of the thing being measured is independent of the measuring process. It is not called into question. The discussion hinges on the way in which the measurement is made, on the 'reliability' of the statistical process, according to models provided by the physical sciences or by industry. In the second case, however, the existence and definition of the object are seen as conventions, subject to debate. The tension between these two points of view—one viewing the objects to be described as real things, and the other as the result of convention in the actual work—has long been a feature in the history of human sciences, of the social uses they are put to, and the debates concerning them. (Desrosières 1998: 1)

We have then a sense that the history of measurement of the social and natural world is woven with tensions between perspectives in which the measure leaves the world untouched and those that see the measure as bringing that world into existence. There are therefore powerful questions both about measures and about the objects under measurement. Are we just measuring, or are we also shaping the thing we are measuring? Our starting point is to think about measurement as being subject to questions about its necessity, efficiency, and form.

As this would suggest, measurement and the controversies around the use of numbers in social life are nothing new in themselves. We have been thinking numerically about the world for a very long time—likewise, we have also been questioning the validity, appropriateness, and cost of such numerical thinking for some considerable period. To think numerically requires an attempt to measure that world as a part of those calculations. In simple terms, it is necessary to find the means of measuring different aspects of that world in order to render more things measurable. Ian Hacking (1990: 5) tracks this lengthy history:

> Galileo taught that God wrote the world in the language of mathematics. To learn to read this language we would have to measure as well as calculate. Yet measurement was long mostly confined to the classical sciences of astronomy, geometry, optics, music, plus the new mechanics…Only

around 1840 did the practice of measurement become fully established. In due course measuring became the only experimental thing to do.

This takes us back to 1840 to find that the practice of measurement was starting to take off as a popular pursuit for creating new types of knowledge about the social and natural worlds—as Porter (1986: 8–9) has noted, before 1890 and for some while after, 'statistical thinking' was 'truly interdisciplinary', with the pioneers being 'widely read generalists'. Indeed, often the social world was approached in the same way as the natural world, with the aim being to measure in order to understand hidden laws and norms. As Foucault observed in relation to his reading of 'archaic Greek history', changes in measurement around debt and justice for the purposes of distribution and political struggle 'ultimately gave rise to a form of justice linked to a knowledge in which truth was posited as visible, ascertainable, measurable, compliant with laws similar to those governing the order of the world' (Foucault 2014: 228). Such thinking, if this reading is correct, stretches back some way and can be seen as an ongoing phenomenon in which the pursuit of certain forms of knowledge-based truth, order, and power can be seen. As Porter has also argued, the objective was to be able to read that mathematical language in which the world had been written. The way to start such a process was through better and more frequent measurement of its properties.

This chapter takes the concept and practices of measurement as its focus. Clearly the study of measurement in the social sciences has a long and detailed history; this is reflected upon here, but the chapter takes a particular direction. In keeping with the book's aims, this chapter focuses specifically upon the connections between measurement and power. It is interested in how measurement has become part of the social world, both in terms of escalating infrastructures but also in terms of the cultural and political shifts that accompany such technological change. The aim of this chapter is to explore the systems of measurement that make *metric power* a possibility. The chapter begins with a set of resources that enable us to put measurement in context by thinking of the rise of a faith or trust in numbers. This is followed by a section that extends this contextual work through a discussion of the politics of number. This section elaborates on some of the key questions that emerge about metrics, measurement, and power. The third section then develops this further through an exploration of work on biometrics. Here biometrics serves as a particular instance in which the measurement of life and the body is taken as political in its connections to biopower. Biometrics is important in this case because it is a concept that further elaborates these connections between power and measurement, whilst also enabling an exploration of what might be thought of as the intensification of measurement in our lives. Metrics here then have both a growing power and a growing presence. The chapter concludes by showing how measurement limits aspects of the social in different ways. It concludes with some reflections on the power of measurement to limit, to define value, and to contain the social, bodily, and human life.

Measurements in Context: Faith, Trust, and Reasoning

Clearly, as I have already suggested, there has been much written in philosophy, political theory, and sociology about measurement and calculation. An overview of such literature would require a good deal of space. Rather than attempting to bring all of this literature together, I instead want to focus on accounts of measurement that open up the conceptual terrain that fits with the aims of this book. In particular, we can take Ian Hacking's powerful argument that numbers, statistics, and classifications are responsible for 'making people up' (Hacking 1990: 3) as being particularly pertinent. As is Porter's (1995: 17) observation that numbers 'create new things and transform the meaning of old ones'. Indeed, Porter (1995: 37) contends that the very 'concept of society was itself in part a statistical construct'. This is what he refers to as the 'creative power of statistics' (Porter 1995: 37). We begin then with the sense that measurement and power can be closely allied and that they need to be understood together. In Porter's work, measurement has the power to be active in making, creating, and constructing. From this starting vantage we immediately see that measurement is not just understood to capture or record but also to be an active constitutional presence.

As this would suggest, in this chapter we will focus on accounts of measurement that in some way carve out the conditions of possibility for the circulation of measures and the production of certain possible outcomes (which are then pursued in Chaps. 3 and 4). We should begin by re-emphasising that the measurement of people and populations has, quite obviously, a very long historical arc. Anyone who has tried to use Ancestry.co.uk to track their family tree, for instance, is likely to have quickly discovered that they are only able to trace their family back to the early nineteenth century or late eighteenth century—often the trail stops there (on the census, see Rose 1991: 686–688). But this is just one visible marker of the traces left by people's lives. In terms of the broader infrastructure of which the increased measurement of people is a part, we might note Mike Featherstone's (2000, 2006) point that modernity was defined by an 'impulse' to archive and by the push to capture lives as 'individuated' singularities. Porter (1995: 77) also suggests that 'social quantification means studying people in classes, abstracting away their individuality'. Featherstone notes the relations between the archive and systems of power, but he also suggests that the archive can now be stretched outwards to incorporate many aspects of our everyday lives (see also Gane and Beer 2008: 71–86). The suggestion here is that the impulse to archive has spread into our lives and that archival forms of social media make it possible to archive the everyday in various and increasingly intense ways (for a discussion of social media archives, see Beer 2013: 40–62). The point here is that the infrastructures of measurement and the archive as a storage facility have been central to the development of the modern era. Archival-type infrastructures have vastly increased in their storage capacities, have become 'unbound' (Gane and Beer 2008) in the types of material they capture, and have even moved into the everyday. In short, the measurement of life can be thought of as being a product of, or facilitated by, the impulse to archive and an escalating archival infrastructure. I discussed the presence of archives and archival impulses in detail in my previous book (Beer 2013: 40–62).

These archives are the means of accumulation, storage, and retrieval for measurements. But let us look here at the measurements themselves and, briefly, at their history.

Referring to an 1835 Parisian scientific report on the use of statistics to understand populations, Ian Hacking (1990: 81) observes:

> Numbers were a fetish, numbers for their own sake. What could be done with them? They were supposed to be a guide to legislation. There was the nascent idea of statistical law, but hardly any statistical inference. Yes, one could conclude that the French are more prone to suicide than the English. Yes, Guerry could invent (almost without knowing it) rank-order statistics to argue that improved education does not counter crime rate. But hardly anyone sensed that a new style of reasoning was in the making.

Hacking's point is that with the use of numbers around this time, you could see, as he puts it, a new style of reasoning emerging, a new way of thinking about people and population that was based upon numbers and statistics (see, for instance, the discussion of population in Elden 2007: 573). Miller and Rose (2008: 65) identify a similar time frame in which, from the eighteenth century, there was an 'accumulation and tabulation of facts about the domain to be governed'. They add that this required some transformation in the infrastructures of the state and that this type of 'government inspires and depends upon a huge labour of inscription which renders reality into calculable form' (Miller and Rose 2008: 65). This returns us to some of the discussion in Chap. 1, and particularly to Foucault's points about the expansion of means of knowing populations in order to govern them. Hacking (1991: 192) in this sense is highlighting what he writes of as 'a certain style of solving practical problems by the collection of data' that is based upon 'a sheer fetishism for numbers'. The notion of fetishism here is interesting; it indicates that numbers and calculations were pursued as a kind of obsession.

This pursuit of numbers contributes not just to how people are measured in different ways, but also how they are categorised and classified. Indeed, 'many of the modern categories by which we think about people and their activities', Hacking (1991: 182) interjects, 'were put in place by an attempt to collect numerical data'. Or, as he further explains, 'when the avalanche of numbers began, classifications multiplied because this was the form of this new kind of discourse' (Hacking 1991: 192). The way people are measured and the categories used have a close relationship. Porter (1995: 36) notes, for instance, that classifying people is particularly 'thorny' (for more on the uncertainties and highly problematic nature of classifying people see Porter 1995: 41–45). Porter (1995: 42) notes that categories that start out as 'highly contingent' and 'weak' can end up being extremely 'resilient'—once these categories become 'official' then they also 'become increasingly real'. Porter (1995: 43) concludes that 'public statistics are able to describe reality partly because they help to define it'. As such, these statistical measures capture the world whilst also creating it; they are a productive presence, and they become realities (see Beer 2015b). The categories used can also pre-empt the measures and can prove to be obdurate over time—they stick (see also Hacking 1991: 183). Much of this early work was done in health, but this new style of reasoning went beyond this domain. What Hacking notes then is that this way of understanding the social world through numbers was not limited to particular fields, but each becomes a representative example of a style of reasoning. What he also makes clear is that the various ways that people are made up by numbers matter, especially in terms of how they are categorised, treated, and judged. Indeed, Espeland and Stevens (2008: 412) point to categories as playing a key role in the way that numbers shape outcomes and become 'reactive'. Their point is that numbers are used to 'create or reinforce' categories; those categories then intervene in the social world they are being used to represent (see also Rottenburg and Merry 2015: 7). As Hacking adds, 'in addition to new kinds of people, there are also statistical meta-concepts of which the most notable is 'normalcy' (Hacking 1991: 183). As numbers and categories are utilised, this enables norms to be cemented and versions of normalcy to be reified against which people can then be judged (for more on the role of big data in generating norms, see Day 2014: 133). As Miller and Rose (2008: 66) have more recently contended, such calculations 'reveal and construct norms…to which evaluations can be attached and upon which interventions can be targeted'. Arriving at a similar conclusion Foucault's position was that statistics:

reveals that the population possesses its own regularities: its death rate, its incidence of disease, its regularities of accidents. Statistics also shows that the population also involves specific, aggregate effects and that these phenomena are irreducible to those of the family: major epidemics, endemic expansions, the spiral of labor and wealth. Statistics [further] shows that, through its movements, its customs, and its activity, population has specific economic effects. Statistics enables the specific phenomena of population to be quantified and thereby reveals that this specificity is irreducible [to the] small framework of the family. (Foucault 2007: 104)

Again, as echoed in Hacking's later work, Foucault finds the power in numbers and statistics. This comes with the types of details that are captured as numbers, and then in how these numbers are used. The power of statistics, in the above passage, is to be found in the illumination of regularities, in the aggregate effect of knowing populations, in the economic linkages that can be made, and in the positioning of the individual within populations or groups. Foucault points to this as a kind of atomising effect in which the family was no longer seen to be the unit of measurement or analysis. Rather, the measurement of individuals, rather than families or households, is crucial in enumerating and analysing populations. Again we see the individuating effect of numbers being used to see and govern populations. Power comes then from being '*in* the *know*' (Miller and Rose 2008: 66; italics in the original). This is governing a population by 'exerting a kind of intellectual mastery over it' (Miller and Rose 2008: 67). This is power through numerical knowledge of the people being governed, their properties, and the patterns of social life.

In his first series of lectures at the Collège de France, in 1970–1971, Foucault drew upon Greek antiquity, and particularly Hesiod's song *Works and Days*, to draw a number of observations about the role of metrics in justice and debt. This reading led Foucault (2013: 133) to this relatively simple observation about the use of measurement to produce norms:

In Hesiod we saw the vague search for a measure: a measure the sense and function of which are still hardly specified since it is a matter of the measure of time, of the calendar of agricultural rituals, of the quantitative and qualitative appraisal of products, and, furthermore, of determining not only the when and the how much, but also the 'neither too much nor too little'.

This, Foucault (2013: 133) adds, crucially, is 'measure as calculation and measure as norm'. This is worth restating—*measure as calculation and measure as norm*. In this formulation, measurement has the dual role of both capturing and setting standards, it records and produces. As things are rendered measurable, so too are norms then more readily established. Norms, for Foucault, can be expressed in their simplest form as the establishment of a balance between too much and too little, based upon the calculations of volume and timing—but this is about the origins of something much bigger for Foucault. Numbers, as we have seen through the historical work of Hacking, can be used then to create new notions of normalcy or to hammer home or reinforce those that already exist (for more on self-reinforcing measurements, see Stigler 1986: 3). Elsewhere, Theodore Porter (1995: 11) similarly concludes that the 'credibility of numbers' is a 'social and moral problem'. Responding to Porter's point Espeland (1997: 1110) explains, the problem is:

> social, because so much work, organization, and discipline are required to make and interpret numbers; moral, because the use of numbers must respond to demands for fairness, accountability and impartiality.

Thus we can see numbers as creating these two sets of closely allied problems, encompassing the social and the moral, which need parallel consideration. Not least through a focus upon numerically established norms. Metrics can be organisational and technical in their appearance but they also present questions of judgement and fairness.

These freshly established norms can be seen to be metric-based rules. For Hacking, the general falling-out of favour of determinist modes of reasoning, which were problematised by the facts that new statistical measures revealed, was accompanied by new social rules of sorts. As Hacking explains:

> The erosion of determinism made little immediate difference to anyone. Few were aware of it. Something else was pervasive and everybody came to

know about it: the enumeration of people and their habits. Society became statistical. A new type of law came into being, analogous to the laws of nature, but pertaining to people. These laws were expressed in terms of probability. They carried with them the connotations of normalcy and of deviations from the norm. (Hacking 1990: 1)

As society became statistical, what it is to be considered normal was established in the numbers. Powerful social facts or rules about people, behaviours, differences, and lifestyles could then be promoted in powerful ways through the use of these numerical accounts of the world—however problematic and unpalatable they may be. The power here is in the way that such indicators are 'typically presented as taken-for-granted facts' (Rottenburg and Merry 2015: 5). This, Hacking suggests, can be thought of as a new kind of law that was based around notions or probability (we will revisit probability later in this chapter and also in Chap. 4). This type of calculation of probability and uncertainty can, according to Desrosières (1998: 16), be traced back in different branches to the 1660s—but it wasn't until the nineteenth century that these different traditions combined. In this sense, it was the 'measurement of uncertainty' (Stigler 1986)—which Porter (1986: 149) refers to as the 'science of uncertainty'—that became of increasing importance with this shift to notions of probability. Stigler's (1986:4) history, for instance, claims that:

Over two centuries from 1700 to 1900, statistics underwent what might be described as simultaneous horizontal and vertical development: horizontal in that the methods spread among disciplines, from astronomy and geodesy, to psychology, to biology, and to the social sciences, being transformed in the process; vertical in that the understanding of the role of probability advanced as the analogy of games of chance gave way to probability models for measurement, leading finally to the introduction of inverse probability and the beginnings of statistical inference.

Clearly this passage, which is useful, in that it gives a very board overview of the development of statistics across two centuries (see also Stigler 1986: 358–359), draws us towards a technical history that we are not going to elaborate upon in this book—largely because it is beyond its

remit but also because it has been done so successfully by those I reference in this chapter. Instead, we might use this passage from Stigler to
give a sense of the development not just of technique but, as we have
already suggested, a history of *a way of seeing the world through measurement*. This is a theme that will resonate throughout the rest of the book.

Hacking understands these developments in terms of a mode of thinking that, as we have seen with Jamie Peck's (2010) work on neoliberalism, is geographically specific. Again using the 1820s–1830s as a point of
departure, Hacking (1990: 33) argues that 'every country was statistical
in its own way'. Italian cities, for instance, Hacking (1990: 16) indicates,
'made elaborate statistical inquiries and reports well before anyone else in
Europe'. That is to say that each country produced and utilised statistics
according to its own mode of approach and reasoning. If we take Britain
as an example, Donald Mackenzie (1981) has shown how the history
of British statistics can, in some forms in the late nineteenth and early
twentieth centuries, be associated with the work of those with an interest
in the study of eugenics and heredity (see also Stigler 1986: 263–297 and
Porter 1986: 270–314, 316–317). This is one stream of statistical work in
Britain that, during that period, advanced statistical methods and practices in a particular direction (see also Desrosières 1998: 104, 105–127).
Porter (1986: 319) also argues that these methods evolved in a particular
context in which there was an interest in questions of variation and in the
study of heredity and evolution. Thus, statistical developments became
entangled with certain political positions and agendas (for an account of
the relations of this movement within the development of sociology in
Britain see also Renwick 2014).

Hacking tracks various paths in the development of this statistical type
of reasoning. Hacking (1990: 34) notes that in Europe, despite some differences, each state 'in its own way, created similar institutions to create
its own public numbers'. He adds to this though that:

> [s]ince different administrations counted different things, the numbers
> that were heaped up differed from case to case. National conceptions of
> statistical data varied, and I argue for important differences between the
> ideas of Prussia and those of France, for example. (Hacking 1990: 34)

So, despite some structural similarities as states developed institutions to gather numbers, there were variations in what was measured and how those measures were used. Indeed, there are still now ongoing and recent calls for 'better' data on social mobility and for an improved and more detailed 'measure of a nation' (see Reeves 2015). It would seem that these early concerns with finding the best measure of populations remain today in the tracking of variation through national statistics.

The geography of measurement, then, varied between and even within nation states. But the core and more general point here is that statistics and numbers are placed at the centre of governance. Elsewhere, in a set of reflections on how we should do the history of statistics, Hacking (1991: 181) claims that:

[s]tatistics has helped form the laws about society and the character of social facts. It has engendered concepts and classifications within the human sciences. Moreover the collection of statistics has created, at least, a great bureaucratic machinery. It may think of itself as providing only information, but it is itself part of the technology of power in a modern state.

For Hacking then, statistics and numbers have been responsible for the formation of laws, for creating social facts and for facilitating the concepts, ideas, and categories that we live by and understand the world through. Measurement here is powerful indeed, not just in terms of any obvious forms of social control but in terms of creating, making, and constructing the parameters of social life. Statistics as a brand of knowledge, for Hacking, may look neutral as it objectively engineers apparently objective facts, a 'superficial neutrality' (Hacking 1991: 184), but it is in fact a profoundly felt technology of power with geographical variability. As Porter (1995: ix) has added, objectivity can seem to be about the 'exclusion of judgment' and the 'struggle against subjectivity', yet for him objectivity is actually about 'a set of strategies for dealing with distance and distrust'. In her review of Porter's book, Espeland (1997: 1107–1108) surmises that distance is crucial to understanding Porter's argument, and it is the 'insertion of distance between numbers and their users' that lends those numbers authority and enables their spread. Here, measurement is a part of the technologies of power as they are established and extended in different social settings and as they come to mitigate against judgement and subjectivity at a distance.

This then points to what might be thought of as the tensions that surround objectivity. As Porter (1995: 3) explains:

'Objectivity' arouses the passions as few other words can. Its presence is evidently required for basic justice, honest government, and true knowledge. But an excess of it crushes individual subjects, demeans minority cultures, devalues artistic creativity, and discredits genuine democratic political participation.

The result, for Porter, is that objectivity is often not defined but is rather used in vague ways in order to 'praise or blame'. Here we see that the notion of objectivity itself, as a concept, becomes a means of legitimising and extending power. It is used potentially to crush, demean, and devalue. It is not just about the measures themselves but about the notions of objectivity that come with them that gives them such power and purchase. Metrics have the scope to appear objective whilst limiting the possibilities for participation and enabling certain power dynamics to thrive. In this sense, Porter (1995: 5) argues, objectivity is both an 'ideal of knowing' and a 'moral value'. Objectivity then is both a form of knowledge about the world and is used to extend certain values. The result, according to Porter's (1995: 8) historical account, is the emergence of a 'faith in objectivity'—a crucial phase that embodies the pursuit of measurement in the modern state. The modern state is based upon a belief in the power of numbers. He continues with this crucial passage:

A decision made by the numbers (…) has at least the appearance of being fair and impersonal. Scientific objectivity thus provides an answer to a moral demand for impartiality and fairness. Quantification is a way of making decisions without seeming to decide. Objectivity lends authority to officials who have very little of their own. (Porter 1995: 8)

The point is that we can begin to see that the power of metrics is to be located in the sense of objective reason and disinterested rationality that they evoke. They give the appearance or manifestation of providing a basis for decision making that is both fair and accurate, that is impartial and unbiased. Decisions informed by numbers are decisions that are taken for us, decisions that make themselves. Or, to frame this within a contemporary vernacular, these are decisions that, because of the numbers, are considered to be 'no brainers'. The objectivity of numbers brings legitimacy and projects authority onto those who use them. Here we have a starting point for seeing how measures translate into power.

Noting at this point that many of the histories of statistics take a somewhat European focus, which means that we are still yet to get the kind of accounts of social and human measurement that unravel the genuine global and non-Eurocentric accounts necessary for more 'connected sociologies' (Bhambra's 2014; there are of course some global histories but these are not readily available or widely accessible, an exception is the history of Indian statistics provided by Ghosh et al. 1993). Keeping that caution in mind as we continue, we can still note that in the case of Britain in an earlier period, Hacking highlights the importance of the role of Sir John Sinclair (see also Porter 1986: 24). Sinclair, Hacking (1990: 27) explains, established the Board of Agriculture in 1793, this board was important because its role was a least 'part statistical'. 'Numerical amateurs', Hacking observes, 'became public administrators'. Sir John Sinclair, Hacking (1990: 27) recounts, was a 'great landowner and a public man, caught up in the vibrant movement for agricultural reform in Scotland, he had been convinced in Europe that facts and numbers were the handmaiden of progress'. This required a radical step-change in the statistical infrastructure of his home country of Scotland. Because of this, and because of the lack of statistical knowledge about Scotland, Sinclair produced a 21-volume *Statistical Account of Scotland* in 1799. He produced this after having started the process of gathering parish-based data from ministers from the Church of Scotland in 1788 after a trip to Europe (see Hacking 1990: 25–29). For Hacking (1990: 27), 'Sinclair was a one man statistical office'. Sinclair moved to England and set up the Board for Agriculture and in so doing became 'part of the evolving British System of official statistics' (Hacking 1990: 28).

We see here, in this story of Sir John Sinclair, the early formation of a kind of metric assemblage (see Burrows 2012), with statistical gatherings about people and places being used to generate statistical understandings of populations. We also see here how what Porter (1986) calls 'statistical thinking' and numerical reasoning became part of the governance of states. This represents a moment in which there was a general increase in the importance of numbers in the social fabric. This though is just one point of origin for the escalation of numerical approaches to governance in just one geographical space. The deployment of numbers in the social world varied between geographical contexts, with Sinclair, as we have seen, being influenced and inspired by his trip to Europe. So despite the shared pursuit of metrics, the deployment of this emergent mode of reasoning varied to fit with the social setting of that nation state. It would seem that, for Hacking (1990), Sinclair's contribution was telling and, in many regards, seminal in the formation of both statistical reasoning and in building up the statistical bureaucracy of Britain (Hacking tracks other national stories in addition to this). When we come to think about metric power then, we will need to keep in mind that it is likely to be as geographically and historically defined as the systems of measurement around which it is based.

What though of this broader style of reasoning to which Hacking has referred? For Hacking (1990), this style of reasoning is based upon the understanding of probability in order to manage and control chance. And there are other places we can look for a discussion of probability and the reduction of risk (see Ewald 1991) or for explorations of probability in relation to the work of seventeenth-century mathematical games of chance or the nineteenth-century work of Quetelet and others on uncertainty (see Stigler 1986: 62–63, 161–163; Mair et al. 2015: 7). The understanding of the way in which probability and chance are approached is seen to be crucial in understanding the history of the development of statistics and measurement. Hacking (1990: 10) calls this the 'taming of chance', by which he means the 'way in which apparently chance or irregular events have been brought under the control of natural or social law'. This was to understand chance through explorations and analyses of probability. The world, it was supposed, was being made to be less 'chancy'. Taming is about 'gradual change', Hacking proposes, that follows what he calls the emergence of probability, which stretches even further back in the seventeenth century (Hacking 1990: 9). Porter talks, for instance, of the rise of 'political arithmetic' and the use of 'social numbers' in state policy from the 1660s. In terms of the development of notions of probability, one important change was the rise of sampling, which, Hacking (1990: 7) suggests, 'required techniques of thinking together with technologies of data collection'. Again we see technique and modes of reasoning operating together. As he then adds, an 'entire style of scientific reasoning has had to evolve' (Hacking 1990: 7). This is scientific reasoning and argument based on a 'veneer of objectivity' (Hacking 1990: 4) that statistics can be used to evoke. Objectivity it would seem is a central component of the persuasive power of numerical measurement, they gleam with the sheen of objectivity.

Using numbers to understand what is probable represents a key moment of transition in Hacking's account. According to Hacking (1990: 4), 'probability is, then, the philosophical success story of the first half of the twentieth century'. Probability, and understanding chance and risk, become a central organising force of modern society. It is claimed that this central preoccupation means that 'probability and statistics crowd in upon us'. Hacking writes, the 'statistics of our pleasures and our vices are relentlessly tabulated…Sports, sex, drink, drugs, travel, sleep, friends—nothing escapes' (Hacking 1990: 4). As Hacking describes here, it would seem that nothing can pull itself from the gravity of the measures of life, everything is measurable and is more than likely be subject to measurement—or at least there is a drive to try to measure and compare. We should note that Hacking's historically informed list of measurable pleasures and vices was written long before social media and smartphones only furthered the reach of measurements. We can reflect on how all of the things Hacking lists here have been measured and captured in new ways by combinations of mobile devices and social media profiles. As such, we can see that rather than being a sudden change, we are looking at something that is a culmination of historical factors. Our pleasures and vices have not only just been measured, but there is certainly a new voracity to those measures today.

An important point here, which refers back to the discussion of Will Davies' work in Chap. 1, is that of comparability. Measurements need to be comparable in order to be useful and to allow for differentiation—with indicators being central to enabling those comparisons to be made (see Rottenburg and Merry 2015: 5). As Stigler (1986: 1) points out, to 'serve the purpose of science the measurements must be susceptible to comparison'.

Comparability is crucial. Thus having standards, consistency, and rules is necessary. The grounds upon which comparisons between outcomes need to be made are thus vital, as are the measures that allow comparisons to be made. Also, as we have seen in Chap. 1, comparability is a necessity if competition is to function. Stigler (1986: 1) adds that the 'comparability of measurements requires some common understanding of their accuracy, some way of measuring and expressing the uncertainty in their values and the inferential statements derived from them'. The means by which measures are made comparable then is the product of a history of the measurement of uncertainty, Stigler explains. Probability is one way in which chance was tamed and measures were made comparable; thus it was crucial to enabling the spread of measurement, the impulse to calculate, and statistical thinking. Andrew Barry (2006: 240) describes this 'development of common measurement standards', which renders them comparable, the establishment of a 'metrological zone'. Within this zone measures can be used to contrast and compare objects, which means that such a zone needs to be established in order for comparisons to be made. Clearly then the ability to make numbers comparable is crucial to their use, and particularly their use in differentiation. Hence Espeland and Stevens (2008: 408) focus on the importance of 'commensuration' (see also Espeland 2015: 59). This is the ability to create the conditions in which the relations between objects can be understood, enabling the transformation of 'difference into quantity'. Their point though is that this is not a passive set of outcomes, but rather that commensuration is a 'process' that 'requires considerable social and intellectual investment' (Espeland and Stevens 2008: 408). Making these metrics comparable then takes work. It is a process that requires an active engagement with how the numbers might be contrasted, rendered comparable, and made to reveal difference. Again, we find that when thinking about the role of something like big data, as we know it today, we are pushed to begin to ask questions about the much longer history of measurement and the means by which metrics are rendered comparable.

We have then the pursuit of comparability and attempts to understand chance and probability, but what of the numbers themselves? The general increase in technologies of data collection led to what Hacking (1990: 5) calls the 'avalanche of numbers' (see also Hacking 1991). This avalanche was created as 'nation-states classified, counted and tabulated their subjects anew' (Hacking 1990: 2)—again we are speaking of a time well before commercial organisations joined in with such data harvesting to foster what Nigel Thrift (2005) described ten years ago as 'knowing capitalism'. As Porter (1986: 11) similarly recognises, the 'great explosion of numbers that made the term statistics indispensable occurred during the 1820s and 1830s'. The avalanche or explosion of numbers, it is particularly worth emphasising, was a product of changing infrastructures of measurement. Indeed, Hacking's (1990: 3) point is that the:

> printing of numbers was a surface effect. Behind it lay new technologies for classifying and enumerating, and new bureaucracies with the authority and continuity to deploy the technology.

There is then a whole industry and infrastructure resting behind the final printed number along with a range of systems and structures that are in place to utilise those numbers (this is something we will pick up on in Chap. 3). Hacking notes elsewhere that it was between 1820 and 1840 that 'there was an exponential increase in the number of numbers that was being published' (Hacking 1991: 186; see also Porter 1986). This, Hacking (1991: 186) proposes, was illustrative of a wide-ranging 'enthusiasm for numbers'. We can imagine the excitement that these new possibilities generated, particularly amongst those who were keen to find the means by which the social world could be measured and governed— not least because it would seem that big data is provoking similar enthusiasms today. As Porter (1995: 78) also argues, 'the first great statistical enthusiasm of the 1820s and 1830s grew out of a commitment to the transparency of numbers'—this political commitment to transparency still plays out today with public data resources such as data.gov and what Clare Birchall has called the 'data-driven transparency model' (the issue of transparency and data.gov is discussed in detail by Birchall 2015; see also Ruppert 2015). The fixation with objectivity can perhaps be seen here in the push towards numerically achieved transparency or making transparent through numbers, which has its own politics.

This 'statistical enthusiasm' was embodied in groups like the private statistical society and the formation in 1833 of the statistical section of the British Association for the Advancement of Science (Porter 1986: 31).

The result, for Hacking (1991: 189), is that from the nineteenth century onwards there is 'almost no domain of human enquiry…left untouched by the events that I call the avalanche of numbers, the erosion of determinism and the taming of chance'. Again, Hacking makes this point about little being untouched by the 'avalanche of numbers' occurring and well before the internet, mobile devices, and social media markedly changed the landscape. And then he also forces us to look back to earlier historical periods to see this 'avalanche of numbers'. It would seem that the notion that we are facing an overwhelming deluge of numbers is certainly not something new, and it is not a feeling that has arrived with the onset of big data and the digital age. The sense that data were getting bigger and that this expansion was met with enthusiasm is part of a trajectory rather than a rupture. Similarly, Porter (1995: 48) notes that by 1860 the trading of wheat had changed so that data were produced about price and production, with the separation of knowledge of trading from the product itself (see also Porter 1986: 3). There appeared, Hacking (1991: 191) suggests, that there was 'almost always a perfectly good self-conscious reason for the vast majority of new countings'. The expansion of measurement, in the form of 'new countings' or 'new numberings', always came with some form of justification and rationale to make it seem necessary, legitimate, or important. The enthusiasm for numbers has led to the increasing measurement of people and to tumbling waterfalls of numbers accumulating in vast pools. The pursuit of the measurement of life and people stretches back historically then, and can be seen, as being about an ongoing 'avalanche of numbers' that has a fairly long historical lineage. The avalanche of numbers to which Hacking refers did not start with digitalisation or with the rise of smartphones, apps, or software, it can be traced back further, as can its enthusiasms, consequences, forms, and categories.

The Politics of Number

Expanding upon the politics of numbers requires us to draw upon some more philosophical resources to compliment the historical sources that we have worked with so far. One particularly useful touchstone is Stuart Elden's (2006) *Speaking Against Number*. Elden's (2006: 2) book uses

Martin Heidegger's work to explore 'the interrelation of number and politics'. Elden's (2006: 11) objective was to think through 'the relation of calculation to the political' and to elaborate upon 'the question of measure'. What he finds is that there are certain conditions of possibility that afford calculative approaches to the social that have, as we have discussed above, a long genealogy. Elden (2006: 2) argues that 'grasping this determination of the world—that to be, is to be calculable—is useful in understanding the modern notion of the 'political' as a whole, not sociologically, empirically, ontically, but *ontologically*'. This is to say that to exist is to be rendered calculable. We exist in the numbers. This is to claim that calculation becomes the basis upon which politics finds its force—it is also suggestive, again, of how these forms of calculation work at the level of the individual life. To exist is to be calculable, suggesting that those things that are not calculable are marginalised or expelled from thought. The politics of numbers can then be revealed if we are to explore this mathematical determination of the way the world is—not just through the measurements and calculations but through the forms of thinking that enable these to be realised.

As I have already indicated in Chap. 1, this means that complex associations are enabled that then facilitate comparison and, potentially, competition. For Elden, drawing upon Heidegger, there are dual processes and tensions between similarity and differentiation in such processes. Measurement is used to differentiate, but at the same time the things that are being measured need to be similar enough in the first place for these measures to operate or be comprehensible. As Elden (2006: 3) explains in this key passage:

> Determining things as different and seeking to render them more equivalent, or counting them the same in the first place, requires a number of important moves: most importantly, recognising things as sufficiently similar in their essence that they can be summed or evaluated against each other. In other words, while examinations of these issues are necessary, what would make them more sufficient is an examination of their conditions of possibility. This would be to examine how mathematics and politics intersect, through an examination of how calculation, the taking of measure, is key to the *constitution* of the modern state. (Elden 2006: 3)

Similar enough to measure, but different enough to compare—this is a formulation that we have already encountered and which we will revisit a number of times in this book. The broader issue is how calculation is a central part of state-making and the constitution of the modern state. But there is a further set of arguments emerging here in which ordering processes are put in place and in which certain comparators are defined, this is where categorisations are used to decide what (or who) will be measured against what. Alongside this, Elden argues that these ordering processes and measures are important in themselves, but that a more insightful analysis would be based upon understanding the 'conditions of possibility' that led to the existence and form of such calculation, measurement, and ordering. As I will go on to argue, we may need to further extend our thinking about these conditions of possibility, focusing not just on the measures but also on the way those measures circulate through the world (as discussed in Chap. 3).

These important arguments in Elden's book draw us to an analysis of politics and number that attempts to understand them not as just technological developments but as deeply political shifts. He further clarifies this insistence on the importance of not just seeing measurement as a product of technological developments:

> Calculation is grounded by the science or knowledge of the mathematical, and is set into power by the machination of technology. This is somewhat ambiguous, and could seem to suggest that calculation is dependent on technology, but the suggestion is the reverse: technology is dependent on calculation, which is grounded in a particular way of thinking the mathematical. Technology merely makes this more apparent. (Elden 2006: 139–140)

In this sense, calculation becomes a term that is not solely about the technologies used but is about the processes of ordering and how these processes and orders are understood. Technology facilitates calculation, according to Elden's accounts, but calculation is actually a way of thinking about the world—which of course echoes those arguments that we have seen from Porter and Hacking. It carries with it a particular view of how the world can be approached. Thus, he concludes, technology also requires calculation. That is to say that technology requires a particular understanding of mathematics and of how the world can be measured. Measurement is not just about technological developments, then; it is also about the way of thinking that affords their integration (see also Elden 2006: 142).

The scope of this is telling, with calculation coming to define the parameters of the social and the political. In this sense, 'the incalculable is only the not yet calculable' (Elden 2006: 140). Everything is either counted or waiting to be counted. The direction here is one of expansion of the technological capacity to measure and the political desire to expand such possibilities for inclusion in the metrics. Or, in Heidegger's own words, 'calculation refuses to let anything appear except what is countable' (Heidegger in Elden 2006: 147; the original passage is in Heidegger 1998: 235). Those things that cannot be counted are rendered invisible, and those that can be counted achieve visibility—thus we begin to see arguments about the role of measurement in the creation of value and the politics of visibility, which will be discussed in further detail in Chap. 4. As such, existence, visibility, value, and importance are likely to be defined by what can be calculated and what is measurable. As the scope of measurement expands, the nature of what is valued and what is calculable are likely to be realigned or reinforced. This does create a small question about whether it is possible that ways of calculating will emerge from the possibilities of the technology rather than the reverse. As such, technology potentially may set the agenda or provoke new types of calculation. We are seeing this with big data, where data by-products are used to explore new possibilities for measuring the social beyond the design and intention of the data being produced (e.g. see Beer 2015b). But Elden's point still pertains, which is that calculation is a way of thinking. The conditions of possibility are created as much by this way of thinking as they are by technological developments. Alongside this, Elden's discussions present further questions about what happens to things that are potentially incalculable; we will return to these in Chap. 4.

The result of all of this is that numbers shape perception and thus carve boundaries into the social world and into the practices of individuals. Elden (2006: 143) notes that:

Our view of the world is therefore not only shaped by our perception, it is also limited by it. The ontological foundation of modern science—this notion of calculation—acts to limit the ontic phenomena it, and we, are able to experience and to encompass.

This contention points towards the links between calculation and possibility that I will discuss in Chap. 4, but for the moment we can see that calculation, as in the work of Foucault, is understood here to create limits of different types and thus shape how we experience those limits. Calculation does not just inform; it creates boundaries, obstacles, and liminal edges. As such, order and calculation can be seen to be deeply entwined. Elden (2006: 147) points out that these 'two are related to each other in that dividing something into elements helps establish control over it, as these can be organised, rendered and further divided, or grouped and forced into similarity'. It is in the role played by calculation in ordering processes that we can begin to see its relation to power and its ability to create limits to knowledge, understanding, and perception.

Yet, it is not just calculation itself that Elden suggests is of political significance here. It is not just the numbers themselves, or the systems that enable them to be drawn-out and utilised. Rather, as we have seen, it is a kind of calculative thinking that is at stake. Elden expands upon the political dimensions of this type of calculative thinking discussed earlier in this chapter. As Elden puts it, 'perhaps number alone is not what needs to be resisted, but the mode of thought that makes possible such mere enumeration' (Elden 2006: 175). This perhaps gives us some early sense about how metric power, as I describe it, might be resisted or subverted. The implementation of measurement and calculation in the social world, bringing its limits and experiences, is a product of a way of thinking—a 'mode of thought'—as much as it is an expansion of any calculative infrastructure. Any political resistance would need then, he points out, to act on that mode of thought rather than the calculations themselves. Elden (2006: 180) concludes that:

[m]athematics is not, then, in itself bad, precisely because it removes itself from any prior ethics or politics, but it is dangerous. As a complement, it can have many uses, but when used alone, when the world is reduced to

numbers, a measure, to what is calculable and laid before us; when humans are summed, aggregated and accounted for; then much remains forgotten, unsaid, concealed.

The crucial part in this passage is in its claim that measurement, calculation, and numbers have the power to force us to overlook aspects of the social world. The visibility created by numbers is narrow even as the scope of measurement expends. Metrics lead to particular 'lines of sight' (Amoore 2013: 93). Hence measurement is powerful not just for what it captures and the way it captures it, it is also powerful because of what it conceals, the things it leaves out, devalues, or ignores. In other words, measurement draws attention to certain things, illuminating them in a very particular light, whilst pulling our gaze away from other aspects of the social and the personal (this is something we will develop in more detail in Chap. 4).

There is then a profound politics of measurement that is becoming increasingly significant with the march of both the infrastructures and modes of thought surrounding 'big data' and digital data analytics. There is a need then to be thinking in terms of visibility, differentiation, and other properties of calculative thinking as we address these apparently new developments in data and metrics. Taking something like the field of biometrics we see these very questions of escalating calculative forces taking centre stage.

The Intensification of the Measurement of Life and the Body

Biometrics is perhaps the area in which measurement has been seen to be a central part of the social world and its power relations. It is a field in which the implications of metrics are perhaps most discussed and most developed (most notably in the work on surveillance and border controls, see e.g. Lyon 2007: 118–135; Muller 2010; Magnet 2011). It is in this field that we might note most readily the intensification of the role of measurement in the social world. In his far-reaching book on biometrics, Joseph Pugliese (2010: 2) defines biometric systems as 'technologies that scan a subject's physiological, chemical or behavioural characteristics in order to verify or authenticate their identity'. Biometrics can be understood then as a 'technology of capture' (Pugliese 2010: 2) or as a technology of identification (Jain et al. 1996b). In an interview that I conducted with Btihaj Ajana for the journal *Theory, Culture and Society*, she suggested that generally 'biometrics is defined as a technology of identification that relies on physical characteristics or behavioural traits to identify or verify the identity of a person' (Ajana in Beer 2014a: 329). However, Ajana admitted in that interview that she finds this 'technical definition' to be 'rather limited'. Instead, she argues that:

> biometrics can be defined as a form of 'biomediation' in terms of both the way in which biometrics refashion older forms of identification such as anthropometry and fingerprinting, and also in the way that the body in biometric processes is doubly mediated by being at once the object and subject of measurement, i.e. that which is being measured as well as what enables the process of identification. (Ajana in Beer 2014a: 329)

Ajana here extends beyond the narrower technical account of biometrics to think of it in terms of its genealogy and the way it entwines into bodily practice (see also van der Ploeg 2003). Similarly, in his comparable genealogical position, Joseph Pugliese is also keen to point out that 'biometrics does not only refer to contemporary, computer-automated technologies', but rather, it is a more general term that 'refers to a cluster of technologies that have all been preoccupied with the measurement of the body in order [to] identify, classify, evaluate and regulate target subjects' (Pugliese 2010: 10): from fingerprinting to facial or hand recognition, DNA, or retina and iris scanning (see Jain et al. 1996a). This chimes then with the influential work of Nikolas Rose (2001, 2007) and his far-reaching arguments about what he famously describes as the 'politics of life itself'. His argument is that there is now a politics that is 'concerned with our growing capacities to control, manage, engineer, reshape, and modulate the very vital capacities of human beings as living creatures' (Rose 2007: 3). This new politics operates at the level of biopolitics; that is to say that it operates at the level of life (see also Verran 2012). There is a perception, Rose (2007: 4) argues, that 'we have experienced a "step-change," a qualitative increase in our capacities to engineer our vitality, our development, our metabolism, our origins, and our brains'. With these new possibilities come new opportunities for power to operate on different and more vital scales. It is with these types of developments that, Rose (2007: 40) contends, 'our very biological life itself has entered the domain of decision and choice; these questions of judgment have become inescapable'. The measures of life then become judgements about our very biology. Crucially, one of the five pathways through which this power might operate, according to Rose (2007: 6), is what he describes as the 'economies of vitality' and it is here that we are perhaps to find how measures of biocapital are used to enable power dynamics to unfold and where judgements of value might be based on quantifications of vitality and life.

Of course, Foucault's discussion of biopolitics and biopower is by now very familiar yet there are some things that are worth reflecting back on to contextualise these discussions. He traces biopower back to the seventeenth century and to the emergence of forms of power that operate on the level of the body so as 'to invest life through and through' (Foucault 1998: 139). This form of power was based upon the 'administration of bodies and the calculated management of life' (Foucault 1998: 140). This kind of approach, the 'power over life', replaced, he argues, the old power over death. There was, as Foucault puts it, 'an explosion of numerous and diverse techniques for achieving the subjugation of bodies and the control of populations, marking the beginning of an era of "biopower"' (Foucault 1998: 140). This biopower, Foucault claims, operated through institutions, on the one hand, and population controls, on the other. It is worth noting here that Foucault (2014: 11–12) reflects on his own changing understanding of power in a lecture from early 1980, and specifically his changing views from his mid-1970s to late 1970s work—so Foucault's own take on what this means for power and governance was actually something of a moving and evolving set of ideas rather than a set of fixed conclusions developed throughout his work.

As we have already discussed in relation to the growth of measurement, Foucault (1998: 140) also talks of population controls in terms of 'the emergence of demography, the evaluation of the relationship between resources and inhabitants, the constructing of tables analysing wealth and its circulation'. Indeed, Foucault's argument is that this type of biopower was central to the development of capitalism—with the control of bodies and the manipulation of notions of population crucial to that particular formation of capitalism and its interests. The result of this emergence was that 'biological existence' was accessible as it 'passed into knowledge's field of control and power's sphere of intervention' (Foucault 1998: 142). According to Foucault (1998: 143), this meant that power could now operate over 'living beings' and could be 'applied at the level of life itself'. To understand this, the concept of biopower worked for Foucault (1998: 143) because it could be used to 'designate what brought life and its mechanisms into the realm of explicit calculations and made knowledge-power an agent of transformation of human life'. These then were emergent forms of calculation of life that enabled new forms of knowledge to be used in powerful ways. This approach to power then led to the pursuit of new types of calculation. As Foucault (1998: 146) puts it, 'it gave rise…to comprehensive measures, statistical assessments, and interventions aimed at the entire social body or at groups taken as a whole'. The origins of the contemporary data assemblage were put in place in the pursuit of the measurement of life and the deployment of biopower. As I have already argued, we can then potentially see new types of metric-based power as being part of this lineage and this desire for knowledge of life and the body based upon calculation and measurement. If power is based on the knowledge of bodies, lives, and populations, then it would make sense that forms of advancing calculation and measurement are likely to be pursued—thus expanding the scope, density, and detail of the knowledge through which power is exercised.

A key way in which biopower operates for Foucault is through the establishment of norms. The 'growing importance assumed by the action of the norm' (Foucault 1998: 144) is seen to be a key feature of the development of biopower. These emergent measures of life enable norms to be established and concretised into a material reality, people can then be judged against these norms. Foucault points out that law and other forms of justice and control do not suddenly become unimportant, but rather that these powerful norms assume a prominent place in the regulation of lives. The apparatus that affords these measurable norms then spreads across different institutions. We will be discussing the relations between measurement, power, and norms later in the book (see Chap. 4), but at this stage it is worth noting Foucault's (1998: 144) observation that a 'normalizing society is the historical outcome of a technology of power centred on life'. The more life is measured the more it is exposed to increasingly powerful norms.

All of this is perhaps now quite familiar in terms of offering an analysis of a kind of corporeal form of power that disciplines and regulates life, and there are now numerous places we could go for an extended discussion and critique of Foucault's work (we have already mentioned Nikolas Rose's work as a key intervention here). It is easy to see why commentators of various types have seen Foucault's work as providing opportunities for understanding power as it acts today through various types of biometrics. There is though some way to go from Foucault's historical observations to a full understanding of contemporary biopolitics, particularly given the changing ways in which we see the calculated management of life to which he referred.

As this indicates, despite the shiny newness of the sci-fi type possibilities that a term like biometrics might provoke us to imagine, a biometric type approach can be tied to the lengthy history of social statistics that we have already encountered. Indeed, Porter (1986: 270–314) tracks biometrical statistics back to the nineteenth century in the work of Karl Pearson and Francis Galton. For Desrosières (1998: 139), it was around the turn of the twentieth century that 'a form of statistics shaped largely by mathematics was born of an alliance between biologists and mathematicians, giving rise to *biometrics*'. As Hacking (1990: 22) adds, referring to the first volume of Foucault's work on the history of sexuality, 'no matter how we take Foucault's polarization, biopolitics in some form has been rampant in western civilization from the eighteenth century or earlier'. To emphasise this point still further, in his book on biometrics, Pugliese (2010: 164) concludes that 'contemporary biometric systems are the culmination of a series of anthropometric technologies that can be genealogically traced back to the early nineteenth century and the historical emergence of biopolitics'. The points of origin might vary, but the consensus is that biometrics and biopolitics have been an integral part of the social world for long periods of time and have been thoroughly embedded in systems of power as a result. But despite this, these accounts almost always lead us too conclude that recent years have seen a drastic ramping up of biometric systems.

Turning more directly to the connections between biometrics and power, Ajana points out, for instance, that there are 'embodied and narrative aspects of identity' that are not often included in governmentality (Ajana 2013: 19). Or, as she put it during the interview I conducted with her:

> Given the growing deployment and wide spread biometrics within various sectors and areas of society…it's very important to examine this technology as a rising mechanism and phenomenon of governance, and understand its myriad political, ontological, social and ethical aspects and implications. (Ajana in Beer 2014a: 330)

Pugliese makes a similar argument by also aligning biometrics with biopower to give it a distinctly political focus. He argues that 'framing biometrics within the conceptual schema of biopolitics will enable the fleshing out of the complex intersection of bodies, subjects, technologies and power and the consequent articulation of the lived effects of biometrics as apparatuses of biopower' (Pugliese 2010: 2). The links to Foucault's work are clear here, with his concept of biopower emerging from his work in the mid-1970s and linking into his work on governmentality (see Gordon 1991: 4–5). Both Ajana and Pugliese (along with others including Amoore 2006) are attempting to use the concept of biopolitics to avoid any kind of objective, technical, or neutral encounter with biometrics. This is because biopolitics, Lemke (2011: 7) explains in his orientation to the concept, 'cannot be separated from the economization of life'—thus sub-concepts such as 'biocapital' and 'biovalue' become useful in tracking such relations.

It is perhaps unsurprising then that Pugliese's (2010: 1) accounts indicate that biometrics and biopolitics are deeply inseparable:

> Biometrics, as a technology of authentication and verification, achieves its signifying status only by being situated within relations of power and disciplinary techniques predicated on individuating, identifying, classifying and distributing the templates of biometrically enrolled subjects across

complex political, social and legal networks. As such, biometrics is a technology firmly enmeshed within relations of biopower.

Biometrics, in this incarnation, are always about power and politics. The power of biometrics is in their capacity to order—in the form of individualising classifications—and for these ordering processes to become central to the functioning of various networks of power. As such, for Pugliese, biometrics are about politics and are an integral part of power structures and orders. As he argues:

> Biometric systems, once situated within this context, function as exemplary technologies of biopower. Emerging from a long and complex history of submitting the body to mathematical measurements in order to determine identificatory attributes always charged with political investments, biometric systems are predicated on the notion that the body can be subject to disciplinary economies of explicit calculations, classification, surveillance and control. (Pugliese 2010: 55)

This then is a story of the calculation and classification of the body. The body is measured enabling it to be identified, attributed, and subjected to power. It is based, we see here, on the notion that the body *can* be measured. A mode of thinking in which the body is measurable is then a prerequisite for the establishment of biometrics.

To link these points back to questions of governance we might recall Colin Gordon's (1991: 44) argument that what 'some critics diagnose as the triumph of auto-consuming narcissism can perhaps be more adequately understood as a part of the managerialization of personal identity and personal relations which accompanies the capitalization of the meaning of life'. Gordon's point encourages us to see the developments in biometrics as part of the self-management of identity as well as being the means of discipline and control. Keeping with such a trajectory, and informed by Foucault's interest in practices, Ajana argues for an engagement with the 'rationales, discourses, meanings, dynamics and narratives involved in biometric processes' (Ajana 2013: 15). Ajana's point is that biometrics are 'polysemic' in the implications they have for 'social sorting'.

A crucial point for Ajana is that with changes in biometrics we have seen a 'remediation of measure'—with the pursuit of measurement being channelled through changing technologies. Indeed, the phrase 'remediation of measure' is itself instructive in imagining how technological developments become a part of these systems of biometrics as a consequence of new political discourses and changing media infrastructures. Ajana's position is that, as a result of new types of technological development and the embedding of certain political positions, we have seen a radical intensification of biometrics—with bodies being measured in greater detail and granularity. As she puts it, it is important:

> to challenge the label of newness that is often stapled on [biometrics] and to draw attention to the fact that the body has for so long been the subject of control, measurement, classification and surveillance. The digitalisation aspect of biometrics has certainly intensified such processes and opened up the body to further dynamics of power and control. (Ajana 2013: 45)

This leaves no doubt then that Ajana's take on biometrics is situated in a history of forms of measurement and political regimes, it is not something that has emerged with the rise of digital data or new media. Indeed, Pugliese (2010: 1) argues that 'biometrics has a long and complex historical genealogy that must be tracked back to the nineteenth century and the concomitant emergence of *biopolitics*' (for more on intensification see Foucault 2008: 28). With biometrics we have a form of metrics that has emerged over time but which has intensified in recent years. The body and life itself can now be measured more often and with enhanced detail. The level, scale, and density of biometric processes have escalated drastically (Lyon 2007: 118). This has consequence then for how the body is understood, how bodies move across borders, and for the types of identification politics that they open up. Ajana has indicated that she could 'only see this intensifying as more data become available through the rise of social networking and mobile technologies and the ever-increasing digitization of work, leisure and daily activities and habits' (Ajana in Beer 2014a: 334). Similarly, Pugliese's (2010: 55) genealogy drew him to conclude that 'the body is now subject to an intensification of instrumentalising techniques and procedures'.

The conclusion we can draw from this is that the technological, political, and cultural conditions have facilitated a growth in the measurement of bodies and life. One illustration of such an intensification comes in the form of what is often referred to as 'big data'. For Ajana (in Beer 2014a: 334), 'Big data analytics promises to take the art of measurement to another level…Big data analytics promises to enhance the techniques of prediction and decision-making that have become an important practice in many fields, organizations and sites of governance'. Here big data and their analytics are seen as a product and means for the acceleration and intensification of systems of measurement—and especially biometric systems. Big data become emblematic then of the growth of measurement in everyday life and the vast by-product data that this produces (see also Beer 2013). We have yet to really see big data in these terms or to explore big data as an extension of the biopolitics of biometrics. With big data we have perhaps so far missed the opportunity to return to questions of biometrics so as to see the implications of big data for the body and the measurement of life that they afford (although one key exception here is the work of Crawford et al. 2015).

Pugliese (2010: 3) observes that the 'terms that frame biometric discourse—"authentication" and "verification"—underline the manner in which biometric technologies transmute a subject's corporeal or behavioural attributes into evidentiary data inscribed within regimes of truth'. As such, biometrics, as a form of measurement, is framed by a particular political rhetoric. Biometrics are presented as making bodies and actions verifiable, they become forms of evidence with certain understandings and truths. The discourse surrounding these systems is revealing and again points to their embeddedness in politics and power. These notions of authentification and verification are often then based upon an understanding of biometrics as objective and reliable forms of knowledge. This, according to Pugliese (2010: 5), 'leads to claims that biometric systems are to be celebrated because they are objective technologies that remove the biases and prejudices of human observers, and thus deliver impartial and unmediated knowledge of their respective objects of inquiry'. Part of the power of these forms of measurement is that they are seen to be impartial and direct in the insights that they offer. The reason that they are so powerful is that they are held up as objective and without bias.

The power of metrics here lies not only in what they do but in how they are understood and the resultant trust or faith that people have in them. Thus we have the broader themes that were discussed earlier in this chapter and in Chap. 1 being projected onto the body through the availability of biometrics. When thinking about the role of metrics, we need to think about their consequences for the body and about the way that metrics become embodied in biometrics of different sorts from the weighing scales discussed by Crawford et al. (2015) to DNA and through to border security systems. A pressing absence in the work on big data is a robust examination of the biometrics that are to be found within these growing data and what this means for conceptions of the body or for how those big data take on biopolitical forms. What might be called big biometrics render the body increasingly measurable and open it up to the exact political dynamics that we have discussed in Chap. 1.

Conclusion

Ranging in scale from the nation state to the individual body, this chapter has explored how the intensification of metrics is based upon a long genealogy of development. The ongoing rise of metrics has also been shown to be a product of both technological and cultural changes in our understanding of the social world and how it might be known. We have seen how notions of objectivity are powerful in the spread of measurement. Alongside this, the chapter discussed how comparability and commensuration are used to create and maintain apparently incontrovertible senses of difference, often through the establishment or maintenance of obdurate norms. Similarly, we have seen how the formation of categories is powerful in how people are understood through metrics. There is a type of thinking or an outlook that resides behind the expansion of social measurement. One key property of these numerical measures is that they are seen to be objective and therefore are considered persuasive and fair (see Espeland and Stevens 2008: 416). These themes, along with others covered in this chapter, will echo through the rest of this book.

One of the problems with writing a chapter about measurement is where we might begin and where we might end. We could explore the history of statistics, for example, which would take us into rich if somewhat dense terrain. We might equally have thought in detail about the relation between social science, mathematics, and the natural sciences—with forms of measurement providing the focal point for understanding disciplinary knowledge and the semi-permeable boundaries of shared understanding. Equally, this could have been a chapter that discussed, in detail, the rise of new types of measures such as those in social media, on mobile devices, apps, and other transactional types of metric. There were plenty of options. Instead of taking any of these I have chosen, as I explained at the start of this chapter, to give some sense of context whilst also trying to tease out the questions of measurement that relate to the enactment of the social, the performative power of numbers, and the potential connection of measurement with both circulation and possibility.

Of course, this chapter has not covered all of the bases. Recently, to take one absence, Mair et al. (2015: 5–6) have argued that even Hacking's excellent work lacks insights into 'statistical practice'. Indeed, they point out that the social sciences have largely been silent on matters of practice in statistics and that case studies are needed that reveal these practices. It is necessary, Mair et al. (2015: 5–6) claim, to 'correct simplified conceptions' of social statisticians by exploring the 'understanding work' that they do. They argue that a focus on practice is needed to understand how statisticians 'put society on display' through their work. Beyond this absence, there is still plenty that could be said about the measurement of the social world. Amongst the many possibilities, this chapter has focused upon the key issues relating to the politics of measure that might then be applied as we focus upon the escalating impulse and opportunity to measure that we see today. What Mair et al.'s (2015) work does allow us to do, even though we cannot address their demands directly here, is to reiterate the need not just to think about the measures, the data being used, but to also reflect on how this is interpreted, the techniques and methods that are used to create insights, and how these insights then become part of the social world. This is something we will develop in Chap. 3.

By looking back across the history of measurement, even in the relatively brief form that I have here, we can see that something like the phenomena known as 'Big Data' needs to be situated in the history of 'statistical thinking', the 'enthusiasm' for calculation, and the 'avalanche' or 'explosion' of social numbers. Yet, we should also note that there is still a discernible intensification of measurement that is going on. Quite simply, the combination of the enthusiasm for numbers has converged with the possibilities of new types of data infrastructures to enable the scope and depth of measurement to increase. As a consequence, our lives are being measured more frequently and in a greater number of ways. We are not suddenly being measured because of the presence of smart phones and the like, but we are being measured more often and in new ways. Indeed, with wearables, social media profiles, and a range of smartphone apps mentioned, we see that people even measure themselves for fun and allow themselves to compete with other users. Combine with this the use of statistics in gaming and the like, we see measures as being a fun part of everyday consumption as well as being state-based accumulations of information about individual lives. Measurement, in multifarious ways and based around variegated agendas, is finding an increasingly significant presence in the social world and is intensifying in this presence.

Alongside a contemplation of the context in which measurement occurs and in which metrics accumulate, of which we will discuss more in the following chapter, we have also begun to see in this chapter how measures are used to project objectivity and legitimise decisions. This gives us a helpful starting point for thinking about the power of metrics. That is to say that this faith or trust in numbers, to use Porter's terminology, is also a key part of understanding the metrics themselves. It is in this faith in the objectivity of numbers that metrics are able to reinforce themselves and spread into different social domains. Alongside this, we have seen in this chapter that numbers carry the power to facilitate judgements and decisions, the power to assess value, worth, and merit, the power to shape the way that the social world is understood, approached, and how the people that constitute it are classified and categorised. Measurement then is not neutral. It is a powerful performative presence in the fabric of the social. We may look back hundreds of years to find the genealogy of this set of developments, but we also need to think about how these measures play out today. We have begun in this chapter to think about the measurement of people, life, and the social, in the next chapter we will consider the way that these metrics circulate through the social world. The power of metrics is not just based upon what is measured and how, it is also reliant upon certain pathways of circulation and how these metrics become part of everyday and organisational lives. It is in these circulations that metrics find an audience and are utilised and instantiated. It is by thinking about how they circulate that we will be able to consider how certain measures become so powerful.

References

Ajana, B. (2013). *Governing through biometrics: The biopolitics of identity*. Basingstoke: Palgrave Macmillan.

Amoore, L. (2006). Biometric borders: Governing mobilities in the war on terror. *Political Geography, 25*(3), 336–351.

Amoore, L. (2013). *The politics of possibility: Risk and security beyond possibility*. Durham, NC: Duke University Press.

Barry, A. (2006). Technological zones. *European Journal of Social Theory, 9*(2), 239–253.

Beer, D. (2013). *Popular culture and new media: The politics of circulation*. Basingstoke: Palgrave Macmillan.

Beer, D. (2014a). The biopolitics of biometrics: An interview with Btihaj Ajana. *Theory Culture and Society, 31*(7/8), 329–336.

Beer, D. (2015b). Productive measures: Culture and measurement in the context of everyday neoliberalism. *Big Data and Society, 2*(1), 1–12.

Bhambra, G. K. (2014). *Connected sociologies*. London: Bloomsbury.

Birchall, C. (2015). 'Data.gov-in-a-box': Delimiting transparency. *European Journal of Social Theory, 18*(2), 185–202.

Burrows, R. (2012). Living with the h-index? Metric assemblage in the contemporary academy. *Sociological Review, 60*(2), 355–372.

Crawford, K., Lingel, J., & Karppi, T. (2015). Our metrics, ourselves: A hundred years of self-tracking from the weight scale to the wrist wearable. *European Journal of Cultural Studies, 18*(4–5), 479–496.

Day, R. E. (2014). *Indexing it all: The subject in the age of documentation, information, and data*. Cambridge, MA: MIT Press.

Desrosières, A. (1998). *The politics of numbers: A history of statistical reasoning*. Cambridge, MA: Harvard University Press.

Elden, S. (2006). *Speaking against number: Heidegger, language and the politics of calculation*. Edinburgh: Edinburgh University Press.

Elden, S. (2007). Governmentality, calculation, territory. *Environment and Planning D: Society and Space, 25*(3), 562–580.

Eisenstein, C. (2015, May 4). The oceans are not worth $24 trillion. *Open Democracy*. Accessed August 20, 2015 from https://www.opendemocracy.net/transformation/charles-eisenstein/oceans-are-not-worth-24-trillion

Espeland, W. N. (1997). Authority by the numbers: Porter on quantification, discretion, and the legitimation of expertise. *Law and Social Inquiry, 22*(4), 1107–1133.

Espeland, W. (2015). Narrating numbers. In R. Rottenburg, S. E. Merry, S. J. Park, & J. Mugler (Eds.), *The world of indicators: The making of governmental knowledge through quantification* (pp. 56–75). Cambridge: Cambridge University Press.

Espeland, W. N., & Stevens, M. L. (2008). A sociology of quantification. *European Journal of Sociology, 49*(3), 401–436.

Ewald, F. (1991). Insurance and risk. In G. Burchill, C. Gordon, & P. Miller (Eds.), *The Foucault effect* (pp. 197–210). Chicago: The University of Chicago Press.

Featherstone, M. (2000). Archiving cultures. *British Journal of Sociology, 51*(1), 168–184.

Featherstone, M. (2006). Archive. *Theory Culture and Society, 23*(2–3), 591–596.

Foucault, M. (1998). *The will to knowledge*: *The history of sexuality* (Vol. 1). London: Penguin.

Foucault, M. (2007). *Security, territory, population: Lectures at the Collège de France 1977–1978*. Basingstoke: Palgrave Macmillan.

Foucault, M. (2008). *The birth of biopolitics: Lectures at the Collège de France 1978–1979*. Basingstoke: Palgrave Macmillan.

Foucault, M. (2013). *Lectures on the will to know: Lectures at the Collège de France 1970–1971 and Oedipal Knowledge*. Basingstoke: Palgrave Macmillan.

Foucault, M. (2014). *On the government of the living: Lectures at the Collège de France 1979–1980*. Basingstoke: Palgrave Macmillan.

Gane, N., & Beer, D. (2008). *New media: The key concepts*. Oxford: Berg.

Ghosh, J. K., Mitra, S. K., & Parthasarathy, K. R. (1993). *Glimpses of India's statistical heritage*. New York: Wiley.

Gordon, C. (1991). Governmental rationality: An introduction. In G. Burchill, C. Gordon, & P. Miller (Eds.), *The Foucault effect* (pp. 1–51). Chicago: The University of Chicago Press.

Hacking, I. (1990). *The taming of chance*. Cambridge: Cambridge University Press.

Hacking, I. (1991). How should we do the history of statistics? In G. Burchill, C. Gordon, & P. Miller (Eds.), *The Foucault effect* (pp. 181–195). Chicago: The University of Chicago Press.

Heidegger, M. (1998). In W. McNeill (Ed.), *Pathways*. Cambridge: Cambridge University Press.

Jain, A. K., Bolle, R., & Pankanti, S. (1996a). *Biometrics: Personal identification in networked society*. New York: Springer.

Jain, A. K., Bolle, R., & Pankanti, S. (1996b). Introduction to biometrics. In A. K. Jain, R. Bolle, & S. Pankanti (Eds.), *Biometrics: Personal identification in networked society*. New York: Springer. pp. 1–42.

Lemke, T. (2011). *Bio-politics: An advanced introduction*. New York: New York University Press.

Lyon, D. (2007). *Surveillance studies: An overview*. Cambridge: Polity Press.

MacKenzie, D. (1981). *Statistics in Britain: The social construction of scientific knowledge*. Edinburgh: Edinburgh University Press.

Magnet, S. A. (2011). *When biometrics fail: Gender, race, and the technology of identity*. Durham, NC: Duke University Press.

Mair, M., Greiffenhagen, C., & Sharrock, W. (2015). Statistical practice: Putting society on display. *Theory, Culture and Society*. Online first. doi: 10.1177/0263276414559058.

Miller, P., & Rose, N. (2008). *Governing the present*. Cambridge: Polity Press.

Muller, B. J. (2010). *Security, risk and the biometric state: Governing borders and bodies*. London: Routledge.

Peck, J. (2010). *Constructions of neoliberal reason*. Oxford: Oxford University Press.

Porter, T. M. (1986). *The rise of statistical thinking 1820–1900*. Princeton, NJ: Princeton University Press.

Porter, T. M. (1995). *Trust in numbers: The pursuit of objectivity in science and public life*. Princeton, NJ: Princeton University Press.

Pugliese, J. (2010). *Biometrics: Bodies, technologies, biopolitics*. London: Routledge.

Reeves, R. (2015). The measure of a nation. *Annals of the American Academy, 657*, 22–26.

Renwick, C. (2014). Evolutionism and British sociology. In J. Holmwood & J. Scott (Eds.), *The Palgrave handbook of sociology in Britain*. Palgrave: Basingstoke.

Rose, N. (1991). Governing by numbers: Figuring out democracy. *Accounting Organization and Society, 16*(7), 673–692.

Rose, N. (2001). The politics of life itself. *Theory Culture and Society, 18*(6), 1–30.

Rose, N. (2007). *The politics of life itself: Biomedicine, power, and subjectivity in the twenty-first century.* Princeton, NJ: Princeton University Press.

Rottenburg, R., & Merry, S. E. (2015). A world of indicators: The making of governmental knowledge through quantification. In R. Rottenburg, S. E. Merry, S. J. Park, & J. Mugler (Eds.), *The world of indicators: The making of governmental knowledge through quantification* (pp. 1–33). Cambridge: Cambridge University Press.

Ruppert, E. (2015). Doing the transparent state: Open government data as performance indicators. In R. Rottenburg, S. E. Merry, S. J. Park, & J. Mugler (Eds.), *The world of indicators: The making of governmental knowledge through quantification* (pp. 127–150). Cambridge: Cambridge University Press.

Stigler, S. M. (1986). *The history of statistics: The measurement of uncertainty before 1900.* Cambridge, MA: Belknap Press of Harvard University Press.

Thrift, N. (2005). *Knowing capitalism.* London: Sage.

van der Ploeg, I. (2003). Biometrics and the body as information: Normative issues of the socio-technical coding of the body. In D. Lyon (Ed.), *Surveillance as social sorting: Privacy, risk and digital discrimination.* London: Routledge.

Verran, H. (2012). The changing lives of measures and values: From centre stage in the fading 'disciplinary' society to pervasive background instrument in the emergent 'control' society. In L. Adkins & C. Lury (Eds.), *Measure and value* (pp. 60–72). Oxford: Wiley-Blackwell.

WWF. (2015). *Ocean assets valued at $24 trillion, but dwindling fast.* Worldwildlife. Accessed November 17, 2015, from http://www.worldwildlife.org/stories/ocean-assets-valued-at-24-trillion-but-dwindling-fast

3

Circulation

I wonder if this book might find its way into some sort of league table. It is possible. If you cite it, for example, it might count towards the university world rankings that are produced every year—or it will feed into my own citation or h-Index scores on one of the various citation platforms. Any mention of it on Twitter or in a blog post will potentially contribute to its altmetric score and its relative impact ranking (see Blackman 2015; Wang 2014). Alternatively, it might be that this book will find its way onto a Google Scholar search based upon a combination of the key search terms you use and how these are interpreted by the ranking algorithms. As such, these rankings, based upon the metrics produced about the book, are likely to shape its reception, its audience, who reads it and when.

Our actions can find their way into rankings in lots of different sorts of ways (see Espeland and Sauder 2007: 5). The important thing to consider, though, is that although lots of our actions are measured and captured, *not all measures are equal*. Some become more visible, more telling, or more consequential than others. To give one illustration, Niels van Doorn (2014) looks at how the online scoring system Klout, which can be used to measure the influence of individual social media users, can be seen to be an archetypal system for capturing human capital. As he

© The Editor(s) (if applicable) and The Author(s) 2016
D. Beer, *Metric Power*, DOI 10.1057/978-1-137-55649-3_3

explains, 'such devices proliferate dynamic feedback loops in which their users get to know themselves and others through fluctuating numerical indices that establish both equivalence and difference' (van Doorn 2014: 368). Similarly, Gerlitz and Lury (2014) explore Klout as being a part of the metric-based participative ordering of value in social media. This is just one commonplace type of use of metric feedback loops to judge value—the measure inevitably shapes the content that is then created, which then reinforces itself in the Klout score. As such, we need to combine any understanding of measurement with an understanding of the means by which those measurements become a part of the social world. In this instance, and extending my earlier work (Beer 2013), I use the notion of circulation to give these feedback loops a descriptive label. In keeping with accounts of complexity and complex systems, these circulations are not necessarily chaotic and disordered 'but a dynamic pattern of escalating feedback loops' (Urry 2003: 34). Circulation, which hints at combinations of order and disorder, seems to capture the way that the data that are produced feeds back into the further production of data. We will reflect on this a little further later in the chapter, especially when considering the 'social life of data' (Beer and Burrows 2013). For the moment we might begin this chapter with the simple argument that measurements circulate into the world in different ways, with some being more visible and powerful than others. This visibility and power are a product and part of how they circulate. As such, we need to take a differentiated view to understanding the circulations that are performed as a part of the power of metrics—a big part of their power is in how they move out into the world. Those pathways to circulation vary and so then does the power of that particular measurement. It is not necessarily the measure itself that has power, but how it is realised and integrated into practices, decisions, and processes. The power of those measurements is located not just in what they record or calculate, but in what then happens to those numbers, how they are used and by whom. Here, we can return again to Theodore Porter (1995: viii), who helpfully suggests that his approach is to 'regard numbers, graphs, and formulas first of all as strategies of communication' (see also Espeland 1997: 1108). In this chapter, we follow suit, but we also extend this to think about how these strategies of communication might establish themselves and how they might be a product of the assemblages of which measurement is a central part.

Picking up on some of the arguments from the previous chapter, we can see how statistical thinking and the faith in numbers to which Porter has referred translate into the pursuit of measurement, whilst also shaping how those measures are then utilised. As Desrosières (1998:3) explains:

> Statistical tools allow the discovery or creation of entities that support our descriptions of the world and the way we act upon it. Of these objects we may say both that they are real and that they have been constructed, once they have been repeated in other assemblages and circulated as such, cut off from their origins—which is after all the fate of numerous products.

Measurements, as Desrosières claims, can get 'cut-off from their origins' as they circulate through the social world. They can take on a life of their own. For Espeland, this is about the relations between narratives and metrics. Espeland's (2015: 57) point is that metrics have the effect of striping or eroding narrative out of the thing that they measure, leaving the possibility for new narratives to be imposed on or deduced from the numbers. This is described as the 'interplay between the erasure and invocation of narratives…in the production and reception of indicators' (Espeland 2015: 57). This is a crucial account of how stories may be silenced or reinscribed through metrics. The stories and narratives are stripped out by data, leaving a vacuum to be filled by new narratives based solely on the data. This also reveals something of how metrics circulate. As Espeland (2015: 56) explains:

> The stripping away of narrative facilitates the circulation and insertion of numbers in new locations and their adaptability in new contexts. But as these new forms of knowledge move out and are re-appropriated or resisted by those being evaluated, they elicit new narratives, new stories about what they mean, how they unfold, if they are fair or unfair, or who made them and why.

Taking the social world back to a story-less husk enables the metrics to circulate, as the opportunity is taken to apply new or ready-packaged meanings upon them. Metrics enable narratives to be selected and then applied; this makes them an attractive presence for those who seek opportunities to narrate. Metrics enable the nuance of the event, occurrence, or action to be replaced by the specifics of the stories told through those numbers. The blank page offered by metrics is therefore one of the features that encourages their circulation. The result is that they become active in the narration and renarration of individuals, groups, and the social world in general.

These sometimes wildly circulating measures, Desrosières claims, enable the creation of entities and also shape how we act upon those discoveries—often in support of the way we see or want to see the world. These measures are constructions, of course, but they become objects with very real outcomes. With this in mind, the approach that Desrosières (1998: 30) takes is to 'follow closely the way in which these objects are made and unmade, introduced into realist and nonrealist rhetorics, to further knowledge and action'. The suggestion is that we look at how these measures are made and unmade, how they justify, promote, or limit actions and choices. The approach that Desrosières takes is to explore the connections between 'description' and 'prescription' (Desrosières 1998: 6) or between what he calls 'there is' and 'we must' (Desrosières 1998: 3). In other words, this is to look at how things are measured but also then to look at what those measures are used to prescribe or to declare. This gives us a starting point, but leaves open the space between description and prescription. This is the space between decisions about what is measured and what those selected measures are then used to decide. In this space, we need to think, I would suggest, about the way that measures circulate into the social world—striping and creating narratives, describing and prescribing actions and behaviours. That is to say that we need to think about the systems, styles of thought, infrastructures, and techniques that enable those *descriptive measures to be translated into prescriptive outcomes.*

Generally, there are two types of circulation that we can highlight when thinking about measurement. In broad terms, we can think about these as being the 'social life of methods' (Savage 2013), on the one hand, and 'the social life of data' on the other (Beer and Burrows 2013). The latter is based on the circulation of the metrics themselves. The former is based upon the circulation of the methods that produce that data—

with methods being a technical part of systems of measurement. In other words, the systems of measurement, the methods, themselves circulate through the social world as they move between sectors and are taken up and deployed in different settings. The point here is that the methods as well as the metrics that these systems produce flow through the social world in different and complex ways. Understanding both is crucial to understanding the relations between measurement and circulation. In this chapter, I will focus more centrally upon the circulation of metrics as a form of data, but we should begin by thinking of the 'social life of methods' as the context in which metrics form, accumulate, and circulate.

The Social Life of Methods

There is something of a burgeoning interest in the social sciences with understanding how methods become a part of the social world that they attempt to depict. This work attempts to understand the trajectory of methods as they are deployed in different forms of knowledge creation. These methods might move between academic disciplines, but there is a more pressing interest in reflecting on how methods also move out into other social spheres—with commercial organisations drawing upon social scientific methods to get to know their customers or to further their understanding of the reception of the services that they provide. The result can be an escalation in commercial forms of sociological work of different types (Burrows and Gane 2006) and a transformation in the jurisdiction of social research (Savage and Burrows 2007). Perhaps an obvious example of this is the use of social scientific methods, like the survey or the focus group, within market research. But there is also the scope to see the interest in analysing big data in the commercial sector as an example of the life that methods take on as they are deployed in different contexts. Referring back to Chap. 1, these are practices typical of organisations that are focussed upon creating a competitive advantage through the accumulation of metric-based knowledge and analytical techniques. As further illustration of this diffusion of methods, it has even been argued that popular cultural forms are laced with social scientific types of techniques and insights (Osborne et al. 2008; Beer and

Burrows 2010). Thus, we see methods themselves spreading across the social world and taking on an active role in different forms of knowledge production.

In an agenda-setting piece on the topic, Mike Savage (2013: 4) describes how the 'social life of methods' is based on 'an increasing interdisciplinary interest in making methods an object of study'. This means that the social role and influence of methods can become the focus of analysis, with the result being that it becomes possible to observe the 'social lives' of these methods—to see how they circulate into different spheres. The result, according to Savage (2013: 5), is that:

> methods can thereby be identified as the very stuff of social life. Social networking sites, audit processes, devices to secure 'transparency', algorithms for financial transactions, surveys, maps, interviews, databases and classifications can be seen as modes of instantiating social relationships and identified as modes of 'making up' society.

Methods, Savage suggests, make up the social world. The examples listed in the above excerpt show the scale with which different methods take on a social life and become a central part in the functioning of the social. This then has some echoes of Foucault's (2002b: 417) assertion that 'the emergence of social science cannot, as you see, be isolated from the rise of this new political rationality and from this new political technology'. Foucault's point, as Savage (2010) has extended more recently, is that the methods of social science are part of the political rationality of knowing and understanding a population. It is through such 'political technologies', Foucault (2002b: 417) argues, that 'we have formed in our societies'. Foucault's angle on this is slightly different from that of Savage though, in that he relates these social scientific methods directly to systems of governance. Savage and his colleagues' more recent writings on this topic are dealing with a much more diverse deployment of methods that are used to know and understand in much more multifarious ways. These are methods that have spread far beyond the social sciences.

Savage (2010) calls this the 'politics of method', which refers to the way that the methods used for understanding the social are also a part of that social world. Savage (2013: 5) adds that in the 'most simple sense, then, the "Social Life of Methods" is a response to the increasing salience of methodological devices'. These devices, it has been added, are 'material and social' (Law and Ruppert 2013: 229) in their presence. These devices bring the social life of methods with them; they are 'shaped by the social', they 'format social relations', and they are 'used opportunistically by social actors' (Law and Ruppert 2013: 239). We can see then that devices like smartphones, with their attendant apps, bring these methods to the inside of our everyday routines, as they respond to our data and make recommendations and provide us with predictive analytics of different types.

Savage points out that such an interest in the role of methods in social life is in some ways a 'familiar move' in its insistence 'that social research methods also have a social life'. The point of such a familiar approach is to make methods into an object that 'is made amenable to critical, political analysis' (Savage 2013: 9). But, he argues, there is scope for using a focus on the 'social life of methods' to challenge intellectual jurisdictions and to explore the relations between theory and method in new ways. Not least is the question of digital devices (for a discussion of the role of devices in the social life of methods, see Law and Ruppert 2013). As Savage (2013: 9) points out, in 'the early 21st century, evidence that methods exercise a profound social significance is easy to find amidst the abundance of digital methods'. Savage's point here is that we need only really to look at contemporary media forms to see digital methods in action, as these devices get to know us and as our profiles are used in various commercial analytics. Despite this abundance of digital methods Savage is wary of any 'epochal' accounts of social change (see also Savage 2009). He is keen to point out that this work should not just be about the digital. Savage suggests instead that explorations of the 'social life of methods' can be critical of a good deal of work on digital methods, and that it 'also needs to appreciate the extent to which methods were implicated in forms of governance in earlier historical periods' (Savage 2013: 11; we have also discussed in Chap. 1).

The social life of methods agenda then is historical in its focus and attempts to understand the part played by methods in the formation, maintenance, and transformation of the social. It is aimed at trying to understand the roles performed by methods and to see how methods are used to generate certain understandings of the social world, which then become a part of that social world. Thus, methods take on a social life. Methods pass between parts of that social world enacting understandings and affording certain perspectives and understandings to be deployed.

There has been some suggestion that the current formulation of the social life of methods is limited by its lack of appreciation of practice. As we have already seen in Chap. 2, Mair et al. (2015) argue that focussing upon the practices of statisticians reveals the different ways in which the social life of methods operate. Their point is that the 'social life of methods' type arguments are restricted by their generic take on what methods are and then by the limited understanding of the details of the practices that enable methods to have a particular social life. Mair et al. (2015: 21) are keen to move beyond 'the social life of methods' as being a purely heuristic formulation. We might say though that there seems little obstacle to these two approaches working together, enabling us to see the social life of methods both near and far. Plus we could also add that the work on the social life of methods emerged from observations made of methods in practice. The crucial point that is shared here though, however we choose to approach it, is that methods are active within the social world as they, to use Mair et al.'s (2015) term, attempt to 'put society on display'. The extra question that is generated by these discussions, though, concerns the different sectors in which these methods are deployed. It is not just about the methods per se, but about the knowledge that they produce and the way that they are deployed to various kinds of commercial and organisational ends. In other words, reflecting on the social life of methods might enable us to see the politics of the very deployment of the methods that are used to produce metrics.

In recent years, this social life of methods has even been seen to create a kind of impending empirical crisis for sociology. The arguments are now well known, but Savage and Burrows (2007) widely cited article effectively suggested that the social life of methods was undermining or challenging the jurisdiction of academic sociology, with commercial companies adopting social research techniques of different sorts (see also Savage 2013: 9). More recently, they have updated that earlier co-authored article in light of the debates on big data (Burrows and Savage 2014). In this more recent piece, which revisits their earlier arguments and explores how they need to be adapted for new types of big data and digital methods, they maintain the point that the social life of methods creates new types of problems for the jurisdiction of sociology and the social sciences. They add though that this also presents new opportunities for seeing the social world in new ways. The question of disciplinary jurisdiction and the opportunities and challenges for social research presented by new types of social data are not of central importance to this book, crucial as they are, but what is important at this juncture is to see methods as a part of the social world that they report on. To see methods as having a social life and that they intervene in how the world is understood and how we might respond. Thus, we need to see how this social life plays out and why certain methods take hold in certain settings. As Burrows and Savage (2014) argue, the social life of methods is only escalating as the promises of big data are explored for commercial and, on occasion, political or academic ends. Here methods continue to diffuse outwards across the social world, leaving us to then consider the life of the methods that become established in certain settings. Tracking the methods that reside behind metrics is then important in fully understanding the agendas behind those measures.

As we have already seen then, methods of measurement are not neutral, they come to impinge upon the thing or person that is being measured. It is not just the data produced by systems of measurement that circulates—although this will be the main focus of this chapter—the very methods of measurement also have a social life of their own. We will discuss circulating metrics in more detail now, but this is to acknowledge that methods themselves—where they are deployed and to what ends—are also something that is mobile and changeable, particularly in what people often, perhaps problematically, think of as an era of big data (see Chap. 2).

The 'Social Life of Data' and the 'Politics of Circulation' Revisited

This brings us to the question of data circulations and, within this broader phenomenon, the circulation of metrics. It has become an accepted motif of the day, perhaps even a cliché, that data about our lives are captured and harvested in multifarious ways. The rise of powerful new media infrastructures has made this escalation of data harvesting possible (as discussed in Chap. 2). These infrastructures have become the backdrop to everyday life and are virtually ubiquitous and inescapable in their scope. We live within them. But this is not really news to most people, the news stories about data theft, supermarket store cards and frequently discussed uses of social media data to target advertising would suggest that there is at least a general awareness of the fact that our actions have the potential to generate data. The recent wide-ranging debate around the new surveillance bill—which was deemed a 'snooper's charter'—is one such illustration of the public dialogue around data extraction (for an example of these news stories, see Brooke 2015). We have a sense that data about us are being extracted, but we are not sure in exactly what form, what they capture, or how they are being used. However, these gaps in our knowledge are now being addressed by the kind of work being done by Helen Kennedy (2016), who has meticulously uncovered the ordinary uses of data mining by various types of public and commercial organisations. The types of insights that Kennedy provides reveal the depth of the detail of the types of data that is being mined whilst also showing the limitations of technique, software and data accessibility that often restrict data mining practices within these organisations.

Despite the type of breakthrough work being done by Kennedy (2016) and a range of others working on the ordinary aspects of data mining (such as Turow et al. 2015), we are still limited in our understandings of the lived realities of data circulation that are to be found within the mundane routines of everyday life. As culture has been remediated by these new digital media platforms in particular, including social media and smartphones, so too the escalation of by-product data has become a possibility. It is through these ordinary engagements with devices and media—in work, in consumption, and in our leisure—that we see vast swathes of data accumulating. We have then the need to understand these circulations on many different scales, ranging from the globally connected economic markets of nation states to the algorithmic selection of the next song to be played to you by your music streaming service. This is the context in which we need to consider the vast complexity of data circulating, and where we need to begin to build up the conceptual imagination necessary to grasp these multi-scalar circulations (as outlined in Chap. 2).

Cleary we will not be able to see everything, but we need some points of reference in order to develop such a project. In my previous book I argued that we need to try to understand the 'politics of circulation' that underpins contemporary culture (see Beer 2013). My argument then was that we need to try to understand how data circulate back into culture, transforming the way culture is produced, disseminated, and consumed. It is not enough to be aware that data about our lives accumulate, we also need to understand how data folds back into our everyday lives in different ways. Culture has always had its circulations—of shared symbols, images, and trends—but these circulations of data and metrics represent an expansion and energising of these circulatory pathways. We need to understand the underlying politics of these circulations of data , and we need to understand how these data are sorted, filtered, and directed. This is no easy task. We are talking here about vast unbundled data assemblages that find their way into the variegated everyday practices of a diverse and dispersed set of people. Imagining what is happening in contemporary culture, which we know is fragmented as well as being deeply decentralised, is an almost unfathomable and overwhelming task. My previous book (Beer 2013) argued that there are some focal points that we can adopt to develop a broader understanding of this politics of circulation. I'd like to very briefly reiterate these to enable us to explore how this framework might be developed and, more importantly, how an enhanced version of the 'politics of circulation' might be used to understand the processes that are central to the flow of measurements that constitute and afford *metric power*.

The argument I made was that we need to begin with the objects and infrastructure through which culture is enacted. These objects and infrastructures, which together form assemblages, enable data to accumulate. Once we have a greater understanding of the systems that afford data accumulation, we can then move towards understanding the archiving of data. Thinking of these as archives forces us to think about how they are organised, how the content is tagged and classified, who the gatekeepers are, and how the content can be searched and retrieved. Using this accumulation and ordering of data as a platform, there are then three ways that we might understand the incorporation of data back into everyday life. First, we need to generate a greater understanding of how algorithms filter data and shape encounters—what Ted Striphas (2015: 395) has described as 'algorithmic culture'. Second, there is a need to get a greater understanding of the way that people are now playing with data. Playing with data is an increasingly common part of cultural participation. APIs are frequently made available to enable these data playgrounds to operate. Indeed, there is even an emergent culture of visualisation with individuals using available data resources to create and share visualisations. Alongside this, data aggregators allow for real-time insights into cultural trends—enabling us to see what is 'happening', 'buzzing' or 'hot' at that moment (for a description see Beer 2012). Finally, we also need to think of the way that the body might be implicated by circulations of data. It would be too easy to get carried away with the power of new devices and new software, but we need to give more attention to the ways in which these devices and data circulations are incorporated into bodily routines.

The focal points I have briefly suggested here are intended to enable us to think about the different components in today's data assemblage, and particularly the circulation of metrics within that context. These different components all play a defining part in the politics of circulation. That is to say that each of these dimensions plays a part in the pathways that data take and define their ultimate destinations. Each of these dimensions contributes to the recursive and recombinant data processes that are active in the social world. This though is only really a starting point for understanding the circulations central to metric power. Taking these arguments from my earlier work as a starting point, I'd like to use the rest of this chapter to develop notions of circulation and to think more generally about how contemporary infrastructures enable the circulation of metrics. As such, the following discussion fleshes out some of the properties of contemporary media that I have briefly outlined here. My point is that what is needed is a renewed engagement with the life that data take on as it swirls into the rapid flows typical of contemporary media. We might then separate out two sets of questions in developing these ideas in relation to metrics. First, we might consider the questions around politics and political economy (see Chap. 1). This would be to think about how identities and ideas rebound between broader political forces and individual actors. Second, we might further consider the materiality of the circulations and the infrastructures that afford them. Both of these directions would be required to operate alongside one another if we are to develop our grasp of the contemporary politics of the circulation of metrics.

Metrics and Communicative Capitalism

As the above indicates, the complexity of the circulation of metrics in contemporary society is vast. But as with the systems of measurement discussed in Chap. 2, we should be thinking of these circulatory systems as being historically contingent. Halpern's history of vision and reason is instructive in this regard. In Halpern's (2014: 194) account circulatory systems become part of a shift to cybernetics that is accompanied by the parallel shift from the analogue to the digital:

> For cyberneticians the problem of analogue or digital, otherwise under-stood as the limits between discrete logic and infinity, the separation between the calculable and the incalculable, the representable and the non-representable, and the differences between subjects and objects, was trans-formed into reconfiguration of memory and storage; a transformation that continues to inform our multiplying fantasies of real-time analytics while massive data storage infrastructures are erected to insure the permanence, and recyclability, of data.

Memory and storage then become the central feature of these systems and afford the capture and circulation of data (see also Halpern 2014: 186). Thus the very way that we imagine and understand data circulations is shaped by these 'discourses of data, beauty, and "smartness"' (Halpern 2014: 5). We are then in need of historically sensitive accounts of the circulation of data and metrics that accounts for the changes in infrastruc-tures. These need to sit alongside cultural understandings of data, how they form and what they make possible—with discourses and perceptions of memory and storage being woven into the infrastructures themselves.

As with something like urban infrastructures, it is the case that media, commercial, and organisational data infrastructures are likely to be impossible to fully 'unbundle'. Steve Graham and Simon Marvin (2001) have spoken of the 'splintering' of urban infrastructures, with the various utility providers and systems making the urban space an environment of nested complexity. In such an environment, there can be no complete understanding of all these different systems and their intermingled material properties. We might extend this to think of a splintering of the media by which metrics circulate. Let us for a moment imagine the systems of measurement and the infrastructures that enable their dissemination to be something akin to the pipework, cabling, wires, drains, connections, and interfaces that make up Graham and Marvin's 'splintering urbanism'—with metrics being pumped around like water, electric, gas, or informational signals to streetlights or traffic lights. We might try to reveal and explore these media infrastructures, yet they are likely to remain too vast and complex to be fully comprehended. As with the complex infrastructures of the metropolis, we also have the complex infrastructures of daily governance—with different measures being compiled in different ways, by different organisations to inform different commercial and other interests. There are many nested systems of measurement coalescing in the spaces of our everyday lives. These stretch from the way that wearable smart watches, smartphones, and apps might be used to quantify our own practices and bodies, to the use of workplace performance measures, to the tracking of consumer behaviours in shops and supermarkets, to the extraction of data about our TV, film, or music purchase, to our use of video games and the like. We have a combination of cultural and infrastructural changes at work here that stand as testament to some broader political ideals about the productivity of measurement (as we have already discussed in Chaps. 1 and 2). The question here though is how these infrastructural and cultural changes afford the circulation of metrics. To do this we need to move beyond the conceptual framework I provided in the previous sub-section, and in my earlier book, to flesh out the *political dimensions* of the politics of circulation.

We can begin to attempt to 'unbundle' these circulatory systems with something as mundane as Facebook. The suggestion here is that the circulation of metrics might be a product of the neoliberal self, with us using social media to produce and reinforce the measures we take of our own lives. The 'good' social media profile, one which can effectively compete for attention, could be understood to be a foundational part of the repertoire of the well trained neoliberal subject. The social media profile becomes then a kind of project of the neoliberal self in which we self-train to produce and circulate ourselves more effectively. Philip Mirowski (2013: 113) has argued that social media such as Facebook:

> forces the participant to construct a 'profile' from a limited repertoire of relatively stereotyped materials, challenging the person to somehow attract 'friends' by tweaking their offerings to stand out from the vast run of the mill. It incorporates subtle algorithms that force participants to regularly change and augment their profiles, thus continuously destabilizing their 'identity', as well as inducing real-time metrics to continuously monitor their accumulated 'friends' and number of 'hits' on their pages. It distils the persona down to a jumble of unexplained tastes and alliances, the mélange of which requires the constant care and management by an entity that bears some tenuous relationship to the person uploaded, but who must maintain an assured distance from it.

This is an important passage that links the broader political economy with the mundane and familiar act of the social media profile update. With social media being used to illustrate the processes of self-training and self-branding in the everyday practices of cultural consumption and production (see Hearn 2008). Thus, social media has a quantified or metric set of properties that Grosser (2014) argues influences our interactions, with both *friendship* and *the self* being reimagined in quantitative terms. Ranking can be seen to play a key part in this type of 'reputation economy' (as argued by Hearn 2010). These social media are potentially illustrative then of the embedded nature of what Mirowski (2013) calls 'everyday neoliberalism' (see also Chap. 1). Here social media profiles shape senses and presentations of identity as they come to be filtered through profile structures and algorithmic feeds. We also see the suggestion here that the metrics produced about users' social media use will feed back into these profile-based identities. It could be said that the social media user is being trained by the metrics that they receive back about their activity, thus shaping future activities—this Tweet or blog post got more reaction in the form of shares, likes and click-throughs, so it is more likely that that behaviour or act of productivity will be imitated in the future. Of course, this is only if these commentaries are accurate, which is something we have yet to fully explore. Although to add a further illustration, Twitter's recent addition of the option to 'View Tweet Activity' now enables users to see detailed metrics for individual Tweets. These metrics provide insights into the reception of that particular Tweet—showing how many times it was seen along with the number of times the viewer clicked on any links, expanded the details, engaged with images, clicked on the Tweeter's profile, chose to follow as a result of the Tweet, and so on. These data can only be seen by the person who posted that particular Tweet, which would indicate that they are intended to provide further opportunities for metrically informed self-training (which is emphasised by the accompanying link offering to help the user to reach a larger audience). With those social media users who wish to increase their understanding of the reception of their own Tweets being encouraged to visit the Twitter analytics dashboard for a month by month breakdown of their social media performance.

Wendy Brown (2015b: 34) also notes this type of everyday performance of the neoliberal subject via social media—which she suggests is part of a set of 'strategic' practices that are designed to enhance the 'self's future value'. This can be understood, to extend Brown's point, as a kind of *future proofing of the self* and the protection of its potential value. Thus, we can begin to see how the complex formation of identity, as being contained and destabilised by social media, might be at least partly influenced by the metrics that are produced and circulated about their own performance and how attractive they are as a commodity that people want to befriend or find out about. The implicit argument here is that we judge ourselves and others through their numerical presence in social media. Twitter presents to us a range of simple metrics about our performance. Our profiles hold and publicly present data about our number of followers, how many people we follow, and the number of Tweets we have generated—we can also see how people's Tweets have faired, in terms of likes and retweets. It seems likely that these will feed into decisions about who to follow or friend—which again takes us back to the role of visual metrics in our understanding of reputation (Hearn 2008).

The publicly visible metrics, along with the more advanced privately available data, inform us on what is productive Twitter usage and again shape how we use this media. And of course, this is in addition to the more detailed use of statistics by these social media organisations in their pursuit of targeted advertising and recommendations for friends/followers you might be interested in. Social media is often replete with metrics that circulate back in and out of their usage—shaping our content and profiles—and training us in how to be a productive social media citizen or in how to make our performance visible to others so that we can be judged. As Gerlitz and Helmond (2013) have discussed, these metrics form into a 'like economy' in which metrics about content and profiles are generated and are used to rank and to discern different types of value. I recently noticed, for example, that the academics' own social media platform academia.edu provides rankings of its users so that academics can see in which percentile they are placed for the level of attention that their profile receives.

Mirowski and Brown are not alone in connecting neoliberal forms of governance with the everyday use of social media. Jodi Dean's influential work is concerned with exploring who it is that might be heard and who might listen to such social media-based circulations. The question she raises is about the way that these circulations appear to breed action and visibility, whilst actually having the effect of silencing and obfuscating. Social media, for Dean, are a central part of what she describes as 'communicative capitalism' (for an overview, see also Hill 2015: 7–9). Dean (2009: 49) provocatively claims that:

[c]ommunicative capitalism strengthens the grip of neoliberalism. Our everyday practices of searching and linking, our communicative acts of discussing and disagreeing, performing and posing, intensify our dependence on the information networks crucial to the financial and corporate dominance of neoliberalism. Communicative capitalism captures our political intervention, formatting them as contributions to its circuits of affect and entertainment—*we feel political, involved, like contributors who really matter.* (Dean 2009: 49)

The type of communicative capitalism that unfolds in social media is associated with the broader ideals of neoliberalism that we discussed in Chap. 1. The circulations of information in social media, according to Dean, make us feel like we are involved and active when we are not. The metrics contribute, it is suggested, to making us feel like we matter, with likes and retweets providing the basis upon which our sense of significance is evidenced. Such communications are not heard and are ineffective, so they simply tighten the grip, as she puts it, of neoliberalism. Dean's point here is that social media can give the impression of providing a voice whilst covering up for the fact that the voice is not heard. Here these circulations of metrics promote communicative capitalism and enable the extended reach of neoliberalism. The very actions that fuel social media, Dean is suggesting, are about communicative forms of capitalism rather than about genuine intervention or democratic interaction. Indeed, her suggestion is that these circulations, in their very existence, capture our actions and extract their value. Whatever the content, social media promotes these forms of communicative capitalism by drawing us into these networks and circulations. We have then a form of communicative circulation that is automated in its functions and which may be understood to annihilate the possibility of being heard through the scale and din of its communicative flows (see also Hill 2015: 10, 66). We again see the sense that circulations are outside of the control of individuals and become part of the structures and properties of the systems through which they flow.

Dean's argument is complex and provocative. At its centre is the idea that circulating information, data, and metrics are not necessarily enlightening or empowering (for a similar argument about empowerment and networks, see Lovink 2011—in Lovink's case though there is also the exacerbating problem of the limiting effects of information overload). Rather these media-based communicative circulations can give the appearance of empowerment and make us feel like we have been heard, whilst actually subverting that empowerment.

We have two things to consider here. The first is the disempowering use of circulating metrics—this would be a concern with the use of metrics as a form of control. But Dean is also suggesting that even where we see these circulations as working for us, they are also undermining the

possibilities for genuine empowerment. Thus, circulations are restrictive
and limiting on both fronts. As Dean (2009:17) explains:

> Expansions in networked communications media reinforce the hegemony
> of democratic rhetoric. Far from de-democratized, the contemporary ideo-
> logical formation of communicative capitalism fetishizes speech, opinion,
> and participation. It embeds us in a mindset wherein the number of friends
> one has on Facebook or MySpace, the number of page-hits one gets on
> one's blog, and the number of videos featured on one's YouTube channel
> are the key markers of success, and details such as duration, depth of com-
> mitment, corporate and financial influence, access to structures of decision-
> making, and the narrowing of political struggle to the standards of
> do-it-yourself entertainment culture become the boring preoccupations of
> baby-boomers stuck in the past.

Dean is obviously not talking directly here about metrics, although
we see her discussing the role of metrics in the power dynamics of social
media communication. It is this broader political importance of circulation that we might dwell upon here. This may then give us the means to
see the circulation of metrics in terms of broader trends in the politics of
circulatory media. This is to say that we might begin to see the circulation and dissemination of metrics as a part of Dean's communicative
capitalism.

Dean, like Mirowski, speaks of the prevailing visibility of the number
of friends, blog page visits, or the number of views and listens as being
indicative of the power of communicative capitalism. These numbers are
a measure of the dominance of voice, but they are also often regarded as
a numerical indicator of who should be listened to or who has the right
to speak—we are judged through such basic numerical values. It would
seem that attempts to resist or to question the metrics to which we are
exposed is likely to get washed away in the maelstrom of communicative capitalism. Social media is unlikely then to give a space in which to
challenge metric power, particularly as it is a space that is shaped through
the power of these metrics—in its algorithmic processes as well as in the
visible stats about its users. If we challenge metrics through social media,
we are simply adding more noise to these circulations, and thus further

eroding our voice. As Dean (2009: 47) explains, 'networked communication technologies materialize democracy as a political form that formats political energies as communicative engagements'. If Dean is correct, then it would seem that social media such as Twitter may give us a space to question the power of metrics, but will actually not have any purchase in terms of actual resistance. Indeed, again, we might be made to feel that we are resisting when actually we are just joining in with the circulations of Dean's communicative capitalism (and for more on social media's limits for resistance, see Skeggs and Yuill 2015). Dean might well agree then with Les Back's (2007) suggestion that our ability to listen has been damaged by contemporary media.

The question this raises is about whether there is space to respond to circulating metrics? Can resistances effectively circulate alongside those numerical judgements and visualisations? Although I should add that visualisations themselves might offer opportunities for resistance. Clearly, these are not questions that we can answer here, yet it is worth dwelling on Dean's arguments in order to explore these questions a little further, particularly as they enable us to connect circulation directly with questions of power and political economy (as discussed in Chap. 1). Dean suggests that there are three 'animating fantasies' of communicative capitalism: 'abundance', 'participation', and 'wholeness' (Dean 2009: 25). In other words, these fantasies tell us that there is a surfeit of information accumulating, that we are able to get easily involved, and that we have at our disposal a complete sense of what is happening in the world. These 'fantasies' enable communicative capitalism to expand and are seductive in keeping us involved. Dean's point is that the results of these fantasies are something different from their promises. It is at this point that we can begin to see the question mark placed over our ability to actually be heard above the din and cacophony of social media. Dean (2009: 24) argues that:

> communicative capitalism is a political-economic formation in which there is talk without response, in which the very practices associated with governance by the people consolidate and support the most brutal inequalities of corporate-controlled capitalism.

Circulations, it would seem, fragment resistance. There is speech without riposte. Social media are presented as monological spaces of communication. As Dean (2009: 22) puts it, 'I refer to this democracy that talks without responding as communicative capitalism' (Dean 2009: 22). Thus, these media circulations sustain inequalities and divisions. The very acts of communication are lost in their own mass. This is about the way in which communication becomes the source of value, and in which the act of speaking is the generator of value; listening is not necessarily valued. Dean (2009: 24) claims that this 'commodification of communication reformats ever more domains of life in terms of the market' and so is at the forefront of the sinking of neoliberalism into the everyday (see also Mirowski 2013). Dean's point is that the 'rhetorics of access, participation, and democracy work ideologically to secure the technological infrastructure of neoliberalism' (Dean 2009: 23). So what is spoken of as being an empowered, democratic, and decentralised media in which we are able to participate on an equal footing is actually an extension of the reach of the properties, values, and desires of neoliberalism (as discussed in Chap. 1). To use Elden's (2006) phrase, 'speaking against number' is likely to go unheard in such a setting—especially as it is likely to go against the prevalent logic of the day whilst also being swallowed by these mass layers of circulating communication.

The result of this speaking without response and the judgements made about the numbers that are found on individual profiles mean that messages are easily mislaid or silenced (Dean 2009: 24). Circulatory media, then, for Dean, erode voice even though they appear to be doing the opposite. The result, we can conclude from this, is that we are also restricted in the possibilities to respond to the circulations of metrics to which we are exposed. Even if we are mobilised to try to respond to the measures about us (or even those about other people or social groups), doing so through such media will only contribute further to the reach and power of the types of communicative capitalism to which Dean refers. This is a bleak perspective, although we also have examples of contemporary media being used to challenge political power (see e.g. Ruppert and Savage 2012). The result is that circulating measures are hard to resist if we simply join in with those communicative flows. These media and contemporary communication 'submerges politics in a deluge of circulating, disintegrated spectacles and opinions' (Dean 2009: 24). Our attempt to counter measures may then get swallowed up by this deluge, particularly as the deluge itself falls within the logic of metric power. This is something we should consider when reflecting on the power of circulating metrics and the spaces and opportunities for resistance, response, and correction (the question of resistance is something we will reflect on a little further in Chap. 5).

We have a vision here of contemporary media as being a lively hive of communicative activity. But what does it mean for the circulation of the metrics themselves? How do they circulate in such a media environment? It is certainly the case that 'rather than being inert or dead, in the contemporary world data is brought into existence as active or "alive"' (Adkins and Lury 2012: 6). The lively spaces of Dean's communicative capitalism are just one form of sparky media in which metrics are to be found or discussed. Through this discussion of Dean's work, and linking to our earlier discussions, we have some complex issues opening up here. On the one hand, we have the circulating metrics. On the other hand, we have the metrics that are produced by communicative capitalism—such as followers, visits, retweets, favourites, likes, and so on (all of which then shape who gets heard and how visible they are). And then, finally, we have the use of these communicative media to circulate responses to the role of metrics in our lives, as we turn to social media to have a voice about the way our lives are shaped by metrics. Here metric power is operating on three fronts. First, measurements about us are circulating in various ways. Second, the media forms we routinely engage with generate metrics about us and shape the volume and amplification of our voice. And, third, we have the potential closing down of what might look like the opportunities for resistance or challenge to metrics presented by social media—our points will get lost into the monologic of social media. On all three fronts then, to aid in our understanding of the power of metrics, we might want to give further consideration to the materiality of these circulations of metrics and to the responses that might be made to them.

The Materiality of Circulating Metrics: Where Numbers Get Embodied

If we just momentarily return to Foucault (2007: 352), he once uttered the following:

> The other specific character of population is that a series of interactions, circular effects, and effects of diffusion takes place between each individual and all the others that mean that there is a spontaneous bond between the individual and the others which is not constituted and willed by the state. Population will be characterized by the law of mechanics of interest.

The point to take from this is that what Foucault refers to as 'the mechanics of interest' shape and define 'circular effects'. There are mechanisms, based upon the agendas of interest, that afford circulations of knowledge and information. We are held together as a population, bound by the presence of calculations about our collective properties. This sense of population is the product of circular recursive processes that enable notions of population to form and be maintained. What is important here though is to think of these circular effects as being shaped by this mechanics of interest. If we look forward a little and begin to think about this in the context of data infrastructures and their understandings, then Halpern (2014: 184) similarly argues that 'feedback appeared as a route to reintroduce reflexivity and perhaps self-awareness into systems'. The very notion of feedback came to be seen to be important in the development of systems designed for extracting data.

In their book on neoliberalism, to which I have already referred in Chap. 1, Dardot and Laval (2013: 263) have suggested that:

> [m]anagement techniques (evaluation, projects, standardization of procedures, decentralization) are supposed to make it possible to objectify the individual's conformity to the behavioural norm expected of him, to evaluate his subjective involvement by means of grids and other recording instruments on the manager's 'control panel', on pain of penalization in his job, wage and career prospects.

Again we need to look beyond the gendered language here, but the point remains important. Circulating metrics can often be about management techniques of various types—not just in the workplace but in the management and self-management of people, consumers, workers, content creators, and citizens more broadly. As we have already seen, these metrics feed back into the world to aid the cultivation and reification of norms and to promote conformity. But the important thing in this particular passage is that these metrics transfer themselves or are interpreted through grids and the manager's 'control panel'. Let us think here about notions of grids and control panels. Indeed, we might think of the control panel as a kind of interface (see Gane and Beer 2008: 53–69) through which metrics are accessed, analysed, and interpreted. These control panels, like other interfaces, are not neutral. They give the user particular accounts of the information that they draw upon. A recent project on 'dashboards' (Bartlett and Tkacz 2014) has taken up this issue so as to explore the role and power of visual engagements with data and the design of the presentation of management information. We might look beyond the control panel's used to manage call centres, as in the example with which I started this book, to think about the types of control panels that are used to analyse social media data (for a description, see Beer 2012) or the control panels that we use to assess our own exercise performance on our smartphone, or the control panel that presents us with interactive Web-based visualisation of changing populations, crime rates, music tastes, social media sentiment, and the like. This gives a sense of the role of data analytics and the data analytics industry in the use of metrics (see Beer 2015b), such control panels often enable an interactive engagement with the data rather than simply presenting them in fixed form, users can often play with categories or variables. Indeed, there are numerous service providers who provide data analytics and visualisation solutions (for a list of examples of these types of companies, see Columbus 2015). In addition to these organisational solutions services, there are lots of examples that we could point to here: beta.tweetolife.com provides visualisations of Twitter data based around chosen keywords and Twistori.com shows flows of content linking to particular phrases; then there are data services such as brandwatch.com that enable more aesthetic engagements with our own data (and for an overview of a project exploring emotional responses to such visualisations, see Kennedy 2015). There are also sites that specialise in curating such visualisations, including informationis-beautiful.net, which, as I write this, features visualisations of gender pay gaps, gender-based medical data, what music streaming services are playing, and common myths. Indeed, any search for social media data visualisations brings up numerous lists of the best places to go to visualise your data or those of your network. There is a materiality at play here that shapes the way that metrics are used and understood. This materiality of the interface is important in understanding how metrics become part of the everyday (for a discussion of the materiality of interfaces, see Gane and Beer 2008: 53–68). Metric power is frequently deployed through control panels, which act as devices for enabling circulating metrics to be accessed and interpreted. The design of the interface will play a significant role in how metrics are accessed, understood, and used.

Indeed, as this would suggest, we should see metrics as being deeply material in their circulations. They capture and shape the materiality of everyday spaces, but their circulations are also grounded in material processes. Take, for example, Donald Mackenzie's (2014) description of the Cermak data centre. Mackenzie's detailed descriptions are extremely revealing about just how material the data infrastructures of contemporary media are (or for a similar description of a Facebook data centre, see Harding 2015). As Mackenzie (2014: 25) observes in this particularly revealing passage:

> A data centre is no more detached from the brute physical world than a printworks was. Cermak (…) is full of stuff. No individual computer server is particularly heavy, but at Cermak there are tens of thousands of them, along with hundreds of miles of cabling, giant generators and transformers, 30,000-gallon tanks of diesel and big power distribution units…As you walk around, though, you get constant reminders of what it takes to keep Cermak cool: huge pipes carrying chilled water; the occasional blast of very cold air.

Here the temperature of the room and the complex tangles of wires give some sense of what lies behind the circulatory processes that define contemporary media. This is an example of how these flows are not detached, but they are highly embedded in the material world in which we live, from these server-laden buildings to the devices that we carry in our pockets. To understand the circulation of metrics is to understand the materialities of its infrastructures. These complex assemblages (see Beer 2013) shape what measures can be taken and where they end up. Again, we are returned to Graham and Marvin's (2001) image of 'splintering urbanism' as a kind of metaphor for understanding data infrastructures, but this time in a more literal form.

The self-managing warehouse might provide another example of this materiality of the circulation of metrics. In these spaces, human actors leave the automated warehouse to run itself. The York-based company Rowntrees has such a warehouse (BBC 2015). In this case, the warehouse distribution centre is referred to simply as 'The Building'. This building handles all of the products produced by the adjacent factory as well as handling the incoming orders, stock placement, retrieval, and distribution. The computerised central control terminal deploys robots to store and then pick the products. No humans are directly involved, other than to oversee the systems. The building uses the available data to manage the stock. Here we can see how metrics combine with advanced computational systems to enable the logistics behind consumer capitalism. It is worth noting that Amazon runs a similar type of automated warehouse system (Knight 2015). Again, this is to see metrics as being a material presence and a facilitator of economic and social activities, with these systems drawing upon data to manage and predict the flow of goods. This is to see metrics as being as much a part of the very infrastructure of contemporary capitalism as lorries, vans, forklift trucks, pallets, buildings, and steel toe-capped boots. The metrics that are produced through distribution then flow back into these systems as future orders are managed.

Then, we should add, there is the very material presence of data mining. Helen Kennedy's (2016) recent book, as I've already mentioned, shows just how embedded and ordinary data mining has become, in both the commercial and public sectors. Her empirical studies reveal the way that data mining is adopted in different settings to promote efficiency and to know customers with greater precisions—she also reveals the everyday limitations of data mining and the limited capacities of those using and developing techniques in different organisations (see also Turow et al.

2015). In a separate collaborative introduction to a special issue on data mining, Andrejevic, Hearn, and Kennedy (2015) argue that data mining is now a well-established cultural presence that has influence far beyond those that we might expect. As they claim:

> Data analytics involve far more than targeted advertising, however: they envision new strategies for forecasting, targeting and decision-making in a growing range of social realms, such as marketing, employment, education, health care, policing, urban planning and epidemiology. They also have the potential to usher in new unaccountable and opaque forms of discrimination and social sorting based not on human-scale narratives but on incomprehensibly large, and continually growing networks of interconnections. (Andrejevic et al. 2015: 379)

Here the scale and range of data mining are outlined as a material reality across various spheres, with the various types of data that are produced through our routine engagements with the social world being drawn upon for various purposes. This also leaves little doubt about the Andrejevic et al.'s view of the power of such data mining practices to reform, cement, and shape the social order—data mining then is a key action in the circulation of metrics. Andrejevic et al. (2015: 381) outline a process in which the data about particular content become more important than the content itself, so with data mining you get a process that they describe as 'metadatification'. It is the metadata that is of value—the data about the data. The answer for this, they contend, is to develop a cultural studies of data mining that does not lose focus on content and discourse, but which also focusses its attention on infrastructures and media assemblages (see also Beer 2013). The important point though is that data mining becomes a material presence in variegated organisations, with data routinely excavated from social media and the like by organisations in the pursuit of understanding customers and stakeholders of different types. We have obviously known about this type of practice for some time now (see e.g. Turow 2006), the point here though is that this is now an ordinary practice that can be located in various types of organisations, even those that are not apparently all that technologically savvy (as described in Kennedy 2016).

To give some illustration of the potential scope data mining in action, we can turn to the YouGov profiler which is available at https://yougov.co.uk/profileslite#/. YouGov is an organisation that captures and records 'opinions' and 'habits'. Its profiler allows you to search for any 'brand, person, or thing'. The results are then displayed for the chosen search term. To give an illustration, searching the term 'sociology' draws upon a sample of 8864 people for whom we can then see a range of characteristics including political orientation, favourite brands, favourite food, and so on. We can see, for example, that according to the YouGov profiler, the average people who like sociology are left-leaning, are female and aged between 18 and24, have less than £125 of spare income a month, like swimming and cats, are compassionate but occasionally neurotic, like AllSaints clothes and the TV shows *Family Guy* and *Outnumbered*, watch TV for 1–5 hours a week and spend 50 hours on the Internet, and the information continues. Of course, such a profile should not be readily accepted, but this gives a sense of the way that data can be mined to create profiles and associations from which inferences can then be drawn. We could then combine this with the way that data is mined in the production of postcode-level profiles on Mosaic (Burrows and Ellison 2004) and the like in order to see how data mining can be deeply embedded in everyday organisational practices, particularly as new types of social media data are so readily available and are relatively easy for companies to harvest and analyse, as Kennedy (2016) has shown.

The material properties of the circulations of metrics might even be said to be finding their way even further into our everyday lives. The recent launch of smart watches, such as the Apple Watch (as discussed in Chap. 1), is emblematic of a creeping connectivity in our lives. With greater possibilities for even more embodied data to be mined (as discussed by Crawford et al. 2015: 480). As we briefly discussed, such devices represent a more corporeal connection to the circulation of metrics. These devices extract and connect us to metrics in more profound and inescapable ways. Metrics flow back and forth between wearable devices like the Apple Watch, the Pebble Watch, Fitbit, and the human body. These devices are presented as being a more sensory connection into our informational environments. As I have already described, the Apple Watch is even advertised as giving its user a more 'haptic' experience. These devices extract data, such as heart rate, location, movement, position, and so on, and then use this to analyse and train the body. These, extending the type of role of smartphone apps like Strava, are exercise and lifestyle devices. They are designed to heighten activity, to make us less sedentary, and to perform a role in making metrics an embodied reality. Here data mining works on the level of the body, extending the reach of biometrics and making data mining more corporeal. Metrics circulate directly back into bodily routines and become a part of our lifestyles. The smart watch is just the most extreme instance of something that is already happening with the ubiquitous smartphone and its various tracking applications. The direction of travel is towards the increasing embedding of metricisation in our lives. These devices suggest activities, recommend lifestyle choices, and guide the self-training subject in how to improve themselves, their bodies, and their everyday experiences (linking to the earlier claims of Rose 1999: 104). The production and feedback of metrics then become intimately connected with bodily routines and the pursuit of personalised forms of consumer capitalism. What we might want to take from this is that metrics are not abstract phenomena, existing somewhere out in the ether. Rather, they are deeply material in their form and they are finding their way increasingly—both in their production and in their analytic usage—to the inside of our everyday lives. There is a sense then of an inescapability of metrics as they embed into the material environment.

Part of the argument here is that metrics become embodied in their material presence and in the way that they provoke us to work on our selves or to bend and shape our practices in response to them (I elaborate further on this both here and in Chaps. 4 and 6). Indeed, Lemke (2011: 120) has argued that 'any analytics of biopolitics must also take into account forms of subjectivation, that is, the manner in which subjects are brought to work on themselves.' Metrics might facilitate this self-work and make it measurable, but it is understanding the devices and systems that makes this possible. The smart watch is an archetypal device of this kind of self-work or, as Foucault (2007: 182) has put it, 'self-mastery'. To understand metrics in the context of our bodies and our everyday routines, we will need to understand further how the device is used to enable metrics of different sorts to guide this self-work. The subjectivation of today is relatively dense in its form, with networked devices now being carried around in the form of watches, phones, tablets, and so on. The opportunities for metric informed self-work have escalated. Seeing this as being a part of the metricised self-training of a more competitive subject might be one angle (see Chap. 1)—particularly where devices and applications are used to allow us to compete with ourselves or with others to be the fittest, fastest, slimmest, healthiest, most productive, most disciplined in terms of our calorie intake, and so on. Indeed, the rise of smartphone apps for tracking aspects of our health is particularly notable (see Lupton 2014).

All of this is to see what is now often referred to as the 'quantified self' (Lupton 2013; Neff 2013; Nafus and Sherman 2014)—which is based upon technologies being used by individuals to capture and track metrics about themselves—as a form of self-training and self-disciplining through metrics. As the quantified self-movement continues to expand, facilitated by cultures of self-measurement and the increasing power of mobile devices, we are likely to continue to see an escalating role of metrics in the performance of our everyday lives. The self-tracking subject helps to extend the reach of metrics and also enables competition to spread into aspects of life that were previously out of reach.

In order to understand this kind of everyday material instantiation of metrics within bodily routines, we cannot though simply stop with an analysis of the device and the body to which it is attached. These bodies and devices are a part of a much broader assemblage. This needs to be considered to understand how metrics are formed and circulate through the social world in order to find their way into bodily routines. These are, after all, networked devices. We see again how these circulations operate on different scales, from the body through to organisational data mining and even in the global flow of products through factories, warehouses, and informational server hubs. Metric circulations have a multi-scalar material presence, even if they appear immaterial in their form.

This takes us to the point at which we need to extend our understanding of the very infrastructures that afford the circulation of metrics. These infrastructures take many forms, from the use of business intelligence within organisations, to the routine cacophonous flows of social media, to the GPS-based tracking of delivery routes for parcel or food deliveries, to the predictive recommendation systems that shape our cultural consumption of TV, film, music, books, and even journal articles—the list goes on. The point is that each of these vast systems needs careful unpicking or unbundling. This does not require one project, but a whole raft of innovative projects. In the above passages of this chapter, I have already given some sense of how such a set of projects might proceed and the types of analytical framework they might use. However, we can extend this still further by reflecting on some of the other shared properties that these systems and infrastructures of metric circulation might have.

Entering the Unknown and Maybe Even the Unknowable

Echoing some of these insights into the complexity of the splintering social fabric, Frank Pasquale (2015) has recently argued that we are living in what he refers to as a 'black box society'—this draws upon the popular science and technology studies (STS) term 'black box' which is used for focussing the attention upon systems and processes that are yet to be illuminated. With some traces of Thrift's (2005) earlier concept of the 'technological unconscious', Pasquale's (2015: 1) term evokes the sense that we know little about the infrastructures and 'increasingly enigmatic technologies' that underpin our lives. The 'incongruity' that Pasquale (2015: 3) centres his book upon concerns the increasing erosion of the privacy of individuals, leading to the escalated protection of the secrecy of commercial organisations. Similarly, Ted Striphas' (2015: 406) view is that 'what is at stake in algorithmic culture is the privatization of process.' Again, we see this kind of black boxing, in which the processes of culture become private, hidden, and often unknown—processes become both privately owned and kept private. Taking on a now common motif as the backdrop for this discussion, Pasquale (2015: 3) observes that behind this is the pattern that we are 'tracked ever more closely by firms and government, we have no clear idea of just how far much of this information can travel, how it is used, or its consequences'. Thus the data mining discussed a moment ago is something that we might know exists but we have little sense of the scale or frequency of its actions. Pasquale's position is that secrecy is central to power dynamics in such a setting. As he puts it, to 'scrutinize others while avoiding scrutiny oneself is one of the most important forms of power' (Pasquale, 2015: 3). Metric power, as we shall discuss further in Chap. 4, is attached closely to visibility, secrecy, and opacity.

Pasquale uses this notion of a 'Black Box Society' to think about the way in which we become visible whilst the data infrastructures in which we live become increasingly invisible—although, as we have seen through the work of Jodi Dean, this visibility is highly circumscribed and does not necessarily equate to empowerment. As Pasquale (2015: 191) explains, 'black boxes embody a paradox of the so-called information age: Data is becoming staggering in its breadth and depth, yet often the information most important to us is out of our reach, available only to insiders'. Hidden in Pasquale's black box society are the means by which our lives are captured, but in which that information is protected by commercial interests. In line with our earlier suggestions of a splintering mediascape, it is perhaps no surprise that Pasquale (2015: 6) concludes that 'deconstructing the black boxes of Big Data isn't easy'. Concealed within these black boxes are all sorts of processes and systems that produce outcomes. So, when it comes to something like recommendations, Pasquale (2015: 5) claims, 'the economic, political, and cultural agendas behind their suggestions are hard to unravel'. We know that we are being recommended something, but we do not know the source, agenda, or means by which that recommendation has been made. Taken more broadly, such predictive systems have individual and collective consequences with organisations using them 'to make important decisions about us and to influence the decisions we make for ourselves' (Pasquale 2015: 4). The role of such systems then is in decision making—the decisions made about us *or* those we make ourselves. Thus in Pasquale's 'black box society', a good deal of power is placed in the hands of these lurking systems, as they come to decide how we should be treated or as they guide us in the choices we should make. Indeed, we have the feeling here that nothing escapes the reach of the black box. Such interests 'crowd out', as Porter (2015: 34) has phrased it, things like discretion and wisdom. Such an account of agency may seem overstated, but Pasquale's argument nonetheless places the question of agency, choice and discretion at the forefront of the analysis. In so doing, it provides us with some questions that need to be answered with regard to the social power that operates within these systems, however hidden or however powerful such a set of automated processes might turn out to be in the making of decisions.

Whereas Pasquale relies on the black box metaphor, Ronald E. Day (2014) chooses to use concepts from library and information studies to conceptualise the circulation of information, data, and metrics. The unknowability that Pasquale outlines becomes something much more structured, familiar, and knowable in Days' work (these questions of a 'knowing public' are extended in Kennedy and Moss 2015). Yet, as we look more closely at both of these large and telling studies of circulatory information, we find that they have much in common even if the discourse makes us feel quite different about the similar systems that they are describing.

Day (2014) suggests that we use the concept of the index and the notion of 'modern documentary tradition' to think about the structures and functioning of contemporary media forms. We have moved into an era, for Day, in which data enables increased indexing and documentation. We are then subject to these renewed and reinvigorated processes—which of course have a long history based in the archive and the archiving of individual lives (for an overview see Featherstone 2000, 2006). As more information is generated, the possibilities for indexing that information escalate as well. The result, for Day, is that the processes and practices of indexing play a powerful role in the make-up of the social world. As Day (2014: ix) argues, 'documentary indexing and indexicality play a major and increasing role in organizing personal and social identity and value and in reorganizing social and political life.' For Day, it is by revisiting the relations between documents (including fragments of documents) and indexes that we might come to understand the way in which data become part of the social world and act upon subjects.

As is the case with a number of other accounts, Pasquale and Day's conceptual visions bring into view the gaping spectre of the algorithm, especially as we arrive at the foothills of these complex self-organising systems. It has been argued that we are even living a kind of 'algorithmic life', in which algorithms play a powerful role on various fronts (Amoore and Piotukh 2016). Algorithms are one dimension of the 'black box' society that has actually started to receive some sustained and critical attention; thus it is where we might begin to unpick the component features of these systems. It is now being suggested that algorithms have profound and far-reaching social powers. For instance, as Cheney-Lipold has argued, these algorithms can reinforce categories, facilitate identity-based forms of control, and provide the means by which a 'soft biopower' can be realised. I've discussed the power of algorithms in detail elsewhere (see Beer 2009, 2013: 63–100), but the figure of the algorithm seems to have become more concrete rather than receding into the fog (for an overview of critical research on algorithms, see Kitchin 2014b). The attention paid to the role of algorithms is increasing as we try to understand the 'generative rules' (Lash 2007) that they afford—these, Parisi (2013: 1) argues, are 'algorithmic architectures' in which algorithms have 'generative capacities'. There is no doubt that in trying to understand the way that metrics circulate, we would need to understand the part played by algorithms in filtering, searching, retrieving, promoting, and prioritising those metrics. If we take another example, what is called 'high frequency trading', in which algorithms frequently take rapid decisions that are central to the trading activities of markets, then Arnoldi (2015) has shown the role of algorithms in the functioning of automated trading. Arnoldi also shows how human agents try to manipulate those algorithmic actions. Again, the algorithm, however it is responded to, is an active presence in global financial markets, shaping how data and finance circulate (as discussed through the example of the 'hack crash' in Karppi and Crawford 2015).

Day's conclusion is similar to Pasquale's, in that these new complex data formations promote opacity and secrecy. This can be understood as 'opacity resulting from complexity' (Pasquale 2015: 103). These systems are so complex that they can't be understood in their entirety, and nor can the consequences of their outcomes. Day (2014: 4) argues, for instance, that:

> [w]ith increasing recursivity, scale, and ubiquity in sociotechnical infra-structures, algorithms and indexes have become both more opaque and more mobile, hiding the logical and psychological assumptions that once

were very clear in traditional top-down and universal classifications and taxonomic structures, as well as in other professional information techniques and technologies. (Day 2014: 4)

It would seem that Day is also noting the presence of a kind of black box here, with recursive algorithmic systems operating in ways that hide the very assumptions and logics that are modelled within them. We have moved beyond, for Day at least, traditional top-down taxonomic structures. Instead these self-organising systems are mobile and contingent. This is something that Scott Lash (2007) began to conceptualise in the early stages of what was then being thought of as the Web 2.0 era, with what he described as new types of 'post-hegemonic' algorithmic rules finding their way into our lives through the media-based self-organising systems with which we engage (for a discussion, see Beer 2009). Indeed, it is now becoming more common to see these emergent infrastructures as moving away from top-down forms of power and towards more decentralised and immanent power formations (Konings 2015: 27; and for an example of such a position, see Mason 2015). For Konings, there are now more complex relations between processes of centralisation and decentralisation. His claim is that the 'diffuse nature of modern power refers not to a process whereby it is levelled out and operates across a flat social field, but to a more paradoxical movement of simultaneous decentralization and centralization whereby power becomes diffused in ways that organically generate points of symbolic concentration' (Konings 2015: 39). The argument here is that centralisation and decentralisation need not be opposing or exclusive processes, but might operate together—creating points of focus or concentration. Any fan of TV talent shows like *The X Factor* will know that this is the case, with the centralised TV show providing the focal point for more decentralised social media-based discussions and sharing, which in turn then feeds into more centralised news media stories about the contestants and their lives.

A number of writers from different disciplines are thinking of the types of power that these circulatory data informed algorithmic infrastructures might wield (see the pieces contained in Amoore and Piotukh 2016). As Pasquale (2015: 8) summarises 'authority is increasingly expressed algorithmically', he adds that 'decisions that used to be based on human reflection are now made automatically.' Indeed, the often-evoked notions of agency and discretion are frequently understood to be subverted or eroded by these changes (for an example of the type of commercial endorsement of 'cognitive machines' as workforce 'talent', see Deloitte 2015: 95–98). In short, we are perceived to have little control of how data, including metrics, might circulate, and as a consequence, we have little control of how they influence and shape the social world that they are a part of. Thus the notion of algorithmically expressed authority is often one in which algorithms take decisions for us—from who should be let through national borders, to how to treat customers, which shares to purchase, who we might follow on Twitter, what we find out about our friends on our chosen social network site, and on to what we might watch on TV or the books we read. These algorithms have the capacity to shape what and who we know. As such, we should not just be focussing on data or metrics but, to emphasise again, we should also be thinking about the means by which they fold back into the social world from which they are extracted. As Pasquale (2015: 22) points out, 'critical decisions are made not on the basis of the data per se, but on the basis of data analyzed *algorithmically*: that is, in calculations coded in computer software'. Challenging data does little if we do not understand the algorithms that shape its selection, destination, and visibility (as I will discuss in more detail in Chap. 4).

For Pasquale, along with Mackenzie (2006) and Kitchin and Dodge (2011), one issue here is that these algorithms are woven with social values that then become part of the world they create or shape. As he puts it, 'the values and prerogatives that the encoded rules enact are hidden within black boxes' (Pasquale 2015: 8). The problem then, for Pasquale, is that the 'black box society' that he refers to is based upon certain values that are coded into algorithms. Not only do we have a limited understanding of the algorithms themselves we are also unclear on how the values that are coded within them are formed or the consequences that they are having. Day (2014: 127) concurs with his suggestion that 'these algorithmic manipulations must take place within, complement, and reinforce larger political economies of exchange of which they are a part.' Thus, as has been argued recently by Introna (2015), algorithms can be seen to be part of broader forms of governance. The important point here is perhaps that these encoded models can be self-reinforcing (Day 2014: 128; see also Kitchin and Dodge 2011). Understanding the exact details of these embedded values is virtually impossible because 'the algorithms involved are kept secret…protected by laws of secrecy and technologies of obfuscation' (Pasquale 2015: 9)—hence the notion of the 'black box society'. It is for this type of reason that methods such as 'media archaeology' (Parikka 2012), which looks at the genealogy of media, attempt to get a glimpse inside this black box so as to locate its workings. And of course then we have approaches inspired by STS that attempt to uncover the technicalities of these systems. These have recently found a space in the form of software studies (see Manovich 2013) and computational culture, in which features such as algorithms become the object of study. These STS-type software studies include a growing body of work exploring the technical and cultural workings of systems such as the Facebook news algorithm (Bucher 2012) to Google's PageRank algorithm (Rieder 2012), for example.

We might foreground some of our later conclusions at this point by pausing to reflect on this obfuscation of the values embedded within the circulation of metrics. This is where attention is needed. It is not enough to reflect on the metrics themselves, we also need to reflect on the properties of the infrastructures that have led to the form, visibility, and mode of dissemination of those metrics. The result, as we have already alluded, is that metrics can take on something of a life of their own as they move across these infrastructures. Pasquale talks, for instance, of 'runaway data' (Pasquale 2015: 26) and 'runaway profiles' (Pasquale 2015: 30) in which data is energised and sparked into life by algorithmic systems. This is important because such lively recursivity leads to new types of consequences. Pasquale (2015: 32) contends that 'runaway data can lead to cascading disadvantages as digital alchemy creates new analog realities'. Thus, these vibrant data have consequence for social divisions, as they spread or promote disadvantage and translate into material realities—with measures then creating and reproducing 'social boundaries' (Espeland and Stevens 2008: 414). Similarly, within this we might think of *runaway metrics*. This would occur where metrics are energised by circulatory systems and algorithmic promotion. These runaway metrics would become visible and take on a life of their own as they are shared, drawn upon, and, as predictive systems, afford the vast snowballing of metrics across social and organisational networks. Metrics can get out of control and can take hold, finding out new audiences and seeping into networks. Again, this is not just about understanding the metrics, it is about understanding how they get noticed—be this in the type of visualisations they are represented through or the viral means by which they move out into networks (see Sampson 2012). Ranging from interesting, surprising, and notable stats circulating on social media to the internal organisational sharing of a visualisation that shows where improvement is needed or which apparently sheds light on some underlying yet invisible problem that then draw attention (a number of other instances are detailed in Beer 2015b)—this is why Espeland and Stevens (2008: 422–423) indicate that we should be considering the aesthetic properties of visualised numbers in understanding their politics. How metrics look and how they are visualised can dictate their impact. In each case, these metrics have the capacity to create realities. The work that is done by visualisations is something that needs to be unpacked in order to understand how and why certain metric-based visualisations take hold and runaway (see Kennedy et al., 2016).

We will reflect further on how metrics might lead to such 'cascading disadvantages' in Chap. 4. Let us consider for the moment the nature of the circulations that might lead to these disadvantages. Other concepts have been used to understand this type of runaway property that data has in these lively algorithmic systems. The use of biological notions of the 'swarm' and 'swarming' represents one such case in point (Parikka 2010: 157, 165). In these accounts the swarm becomes a kind of visual metaphor for the combination of order, disorder, and patterning that comes with complex interactive systems. It has also been suggested that what we end up with is a culture in which information spreads around networks like a contagion with viral properties. The phrase 'gone viral' may now have moved out of fashion, but the notion of contagion has retained some purchase in academic accounts of information flows. The vision here is of these types of out-of-control information spreading across the social world, with likely unpredictable consequences and outcomes. Sampson (2012), for example, notes the types of resistance that occur to these contagions, which can take the form of active disconnection. Whether this is possible or desirable is one thing, but the vision of circulation that we often find is one of rapid and uncontrollable circulation that is in some way vital, it is alive or lively (see Kember and Zylinska 2012). Data take on a life of their own as they spread through networks. This is only a vision though, and there is plenty to suggest that the contagion works around the fixed architectures of these infrastructures—their 'contagious architecture' (Parisi 2013)—and finds blockages and problems of various types. The problem that this presents is how we understand such viral spreading and how we conceptualise and provide empirical insights into runaway or contagious metrics. The other question is whether we should accept such metaphors in attempting to conceptualise these circulations in the first place. The problem of understanding the vitality of the circulation of metrics—which itself is likely to be hard given their likely speed and mobility—then becomes a question of developing methods and a conceptual vocabulary that make the circulations visible and which are also able to account for the properties of the infrastructures that afford them.

Conclusion

The above accounts give some indication of how the circulation of metrics shape what is known about the social world. The pursuit of feedback and the integration of recursive systems have been crucial in realising *metric power*. Halpern's (2014: 84) argument here is that 'as the nature of the observer was reconceived, knowledge claims were also transformed.' This is to say that the circulatory systems to which people and populations are exposed have the capacity to transform both the observer and the observed. The result is that knowledge of the social world is also altered. Halpern (2014: 84) continues:

> As cognition, perception, and the body (both social and individual) came to be redefined in terms of feedback and patterned interactions *between* objects and subjects (as a communication process), what it meant to produce a truthful account of the world (or a product) shifted, coming to be no longer about hidden truths, invisible elements, or psychological depths but rather about affect and behaviour.

The very notion of truth then, according to Halpern's historical account, can be transformed by circulatory systems of measurement. With feedback loops of this type, the move is towards the manipulation of bodily and emotive experiences (see Chap. 6 for further discussion of metrics, the body, and emotion). We find, for Halpern (2014: 27), a 'reformulation of vision' in which things are seen differently. Halpern (2014: 26) sees this is a 'historical reorganization of vision and reason' that dates to the mid-twentieth century and which was based on these new infrastructures of 'sense', 'knowledge', and data circulation. This is something we will pick up again in Chap. 6.

At this point, we might be hearing some distant echoes of Michel Foucault's lectures on biopolitics and governance, which we dwelt on a little in Chap. 1. Circulating metrics are the means by which markets can be realised and where objective knowledge enables the social world to fall into place. Metrics become indisputable truths, selected to produce apparently justified inequalities and to enable competitiveness through 'scientific forms of knowledge' (Foucault 2007: 350–351). In such a vision, the power of metrics is in how they circulate through the world and in how they become the 'rational principles and forms of calculation specific to an art of government' (Foucault 2007: 348). This chapter has shown though that metrics circulate in particular ways. To understand the power of metrics is as much to understand these circulations as it is to understand the measurements themselves. The very infrastructures of measurement not only count, they also disseminate. This dissemination can be charged with all sorts of agendas—and may even be the product of the models used to produce algorithmic decision making that sorts, filters, prioritises, and makes visible certain metrics. The circulatory dissemination systems of what Pasquale calls the 'black box society' may be hard to unpick and may have splintered to levels of complexity that are beyond comprehension, but we cannot leave our analysis here. We need to begin to look inside these circulatory systems to see how they work and to see how metrics find their audience. Simply looking at metrics, numbers, and calculations is not enough for understanding their power. Much more than that is needed. Central to this is to examine how metrics become a part of that social world. It is crucial to explore how they are embodied or institutionalised. These discussions give us a frame of reference and conceptual set of resources for pursuing these questions further and for expanding our understanding of the way that certain measures become a part of the social world.

What is at stake here, as Foucault (2014: 6) has put it, are the relations between the 'manifestation of the truth and the exercise of power'. For the 'exercise of power', Foucault (2014: 6) explains, 'is almost always accompanied by a manifestation of truth'. In the circulation of metrics, we find truths being selected and then exercised. He attempts to capture this set of relations using the concept of 'alethurgy'. Foucault (2014: 7) develops this idea in his 1980 lecture series:

> We could call 'alethurgy' the manifestation of truth as the set of possible verbal or non-verbal procedures by which one brings to light what is laid down as true as opposed to false, hidden, inexpressible, unforeseeable, or forgotten, and say that there is no exercise of power without something like an alethurgy.

This concept of alethurgy is intended to explore the way that truth is created in order for power to be deployed. He adds that alethurgy would need to come before something like hegemony, with established notions of truth enabling ideology. The possibility then is for alethurgy, and the relations between truth and power, to be reworked. There can be a 'calculating reorganization' of alethurgy peculiar to particular aims and 'exercises of power' (Foucault 2014: 9). This is to think of the 'notion of government by truth' (Foucault 2014: 11; Foucault explains that this takes us beyond dominant ideologies and beyond his earlier formulations of knowledge-power). The association here is between the 'art of government', discussed in his lectures of the late 1970s, and the 'game of truth' (Foucault 2014: 13). These relations between the 'exercise of power' and the 'manifestation of truth' are, Foucault (2014: 13–16) argues, based around 'exact' and 'specialised' knowledge. Hence Foucault (2014: 93) speaks in those lecture of 'regimes of truth'. Metrics, given the discussion in the previous pages of this book, could well be central to the manifestation of truth that then plays out in the exercise of power. We may need to think then of circulating metrics in terms of alethurgy and truth making. Metrics can be seen to lay down what is true and what is false, and to prescribe what is then hidden, forgotten, or considered inexpressible. Metrics may then be central to the alethurgy of the age and to the organisation of the relations between truth making and power deployment. Metrics are a means, as we have seen, for generating truths.

One way of developing this further, allowing us to move towards an analysis of the reception of these circulating metrics, is provided by Day's (2014: 137) argument that 'in social big data we are not just documentary subjects, not just documentary objects, but rather we are the two conjoined with each other as parametrically viewed historical expressions'. As conjoined subjects and objects, we are active in producing and reacting to the metrics to which we are exposed. Rottenburg and Merry (2015) similarly contend that 'despite the fact that they are designed to produce scientific objectivity, forms of quantification never simply reflect the world.' Espeland and Sauder (2007) recover the methodological concept of 'reactivity' in order to conceptualise the ways in which people respond to the measures to which they are exposed. They use this concept to explore 'the idea that people change their behaviour in reaction to being evaluated, observed, or measured' (Espeland and Sauder 2007: 1). Espeland and Sauder's (2007) suggestion is that we explore 'social measures' through an analysis of both the 'mechanisms' and 'effects' of that reactivity. This concept of reactivity is useful here for a number of reasons, not least because of its focus on understanding the way that measures are incorporated into practice and behaviours. This is a concept then that is designed to shift attention towards the way that metrics provoke as well as capture, produce as well as record (although I will suggest an alternative position in Chap. 6). The point Espeland and Sauder (2007: 7) make is that it is important to understand the 'sense making' processes in measurement. Their point is that by focussing upon 'reactivity', we have to then pay attention to how people interpret and find meaning in measures. It is this attachment of meanings and interpretations which then inform and shape behaviour (I'll return to this in Chap. 6).

In more general terms, we are captured as objects in innumerable measurements. We are then subject to the outcomes that those measurements produce. We respond, we react, we are provoked or stimulated into action, we are inhibited and cowed into inactivity—the list goes on. It is for this reason that Day (2014: 137) sees social big data as 'constituting a form of documentary governmentality'. These data are acting upon us in different ways. This means that we cannot think of metrics or even of circulating metrics in isolation, we need to also think about how they govern us, how they shape our lives, how they do not just measure but also create possibilities as we learn to live with the metrics that circulate through organisations, markets, structures, rankings, and our lives. These are recursive processes.

Of course, as this would suggest, when attempting to understand the role of metrics in the social world, it is important to think not just of those measures but also about how they are 'instantiated' (Hayles 1999) in practices, processes, and routines. That is to say that it is not just a case of what is measured and how, it is a case of how these measures become visible, get noticed, and move out into the social world. In short then, it is about bringing notions of measurement together with an understanding of the 'social life of methods' (Savage 2013) and the 'social life of data' (Beer and Burrows 2013). This is to see how the methods of measurement are adopted and then to see how the metrics they produce circulate into the social world. Metrics only really become powerful when they are acknowledged and prioritised. Some measures disappear; others become the linchpin of how we are judged. As such, we need to understand these circulations of metrics in order to understand their potential power. To understand these circulations, we need to understand the attitudes and imaginaries that afford them and the hierarchy of favoured measures. But we also need to think about the infrastructures that direct the circulation of these metrics. Understanding metric power is about understanding the conditions and attitudes of measurement whilst also understanding the way that preferences and material conditions generate the pathways to their realisation and visibility. Once we have reached such a position, we will be able to see how metrics shape what is possible *and* what is seen to be possible.

References

Adkins, L., & Lury, C. (2012). Introduction: Special measures. In L. Adkins & C. Lury (Eds.), *Measure and value* (pp. 5–23). Oxford: Wiley-Blackwell.

Amoore, L., & Piotukh, V. (Eds.). (2016). *Algorithmic life*: *Calculative devices in the age of big data*. London: Routledge.

Andrejevic, M., Hearn, A., & Kennedy, H. (2015). Cultural studies of data mining: Introduction. *European Journal of Cultural Studies, 18*(4–5), 379–394.

Arnoldi, J. (2015). Computer algorithms, market manipulation and the institutionalization of high frequency trading. *Theory, Culture and Society*. Online first. doi 10.1177/0263276414566642.

Back, L. (2007). *The art of listening*. Oxford: Berg.

Bartlett, J., & Tkacz, N. (2014). Keeping an eye on the dashboard. *Demos Quarterly*, (4) Autmun. Accessed July 6, 2015, from http://quarterly.demos.co.uk/article/issue-4/keeping-an-eye-on-the-dashboard/

BBC. (2015, May 6). Inside the factory: How our favourite foods are made. *BBC 2*. Originally broadcast. http://www.bbc.co.uk/programmes/b05tynw0

Beer, D. (2009). Power through the algorithm? Participatory web cultures and the technological unconscious. *New Media and Society, 11*(6), 985–1002.

Beer, D. (2012). Using social media data aggregators to so social research. *Sociological Research Online, 17*(3), 10 .Accessed July 6, 2015, from http://www.socresonline.org.uk/17/3/10.html?buffer_share=e5

Beer, D. (2013). *Popular culture and new media: The politics of circulation*. Basingstoke: Palgrave Macmillan.

Beer, D. (2015b). Productive measures: Culture and measurement in the context of everyday neoliberalism. *Big Data and Society, 2*(1), 1–12.

Beer, D., & Burrows, R. (2010). The sociological imagination as popular culture. In J. Burnett, S. Jeffers, & G. Thomas (Eds.), *New social connections: Sociology's subjects and objects* (pp. 233–252). Basingstoke: Palgrave Macmillan.

Beer, D., & Burrows, R. (2013). Popular culture, digital archives and the new social life of data. *Theory Culture and Society, 30*(4), 47–71.

Blackman, L. (2015). Social media and the politics of small data: Post publication peer review and academic value. *Theory, Culture and Society*. Online first. doi: 10.1177/0263276415590002.

Brooke, H. (2015, November 8). The snooper's charter makes George Orwell look lacking in vision. *The Guardian*. Accessed November 17, 2015, from http://www.theguardian.com/commentisfree/2015/nov/08/surveillance-bill-snoopers-charter-george-orwell

Brown, W. (2015b). *Undoing the Demos: Neoliberalism's stealth revolution*. New York: Zone Books.

Bucher, T. (2012). Want to be on top? Algorithmic power and the threat of invisibility on Facebook. *New Media & Society 14*(7), 1164–1180.

Burrows, R., & Ellison, N. (2004). Sorting places out? Towards a social politics of neighbourhood informatization. *Information Communication and Society, 7*(3), 321–336.

Burrows, R., & Gane, N. (2006). Geodemographics, software and class. *Sociology, 40*(5), 793–812.

Burrows, R., & Savage, M. (2014). After the crisis? Big data and the methodological challenges of empirical sociology. *Big Data and Society, 1*(1), 1–6.

Columbus, L. (2015, May 9). The best big data and business analytics companies to work for in 2015. *Forbes*. Accessed November 2, 2015, from http://www.forbes.com/sites/louiscolumbus/2015/05/09/the-best-big-data-and-business-analytics-companies-to-work-for-in-2015/

Crawford, K., Lingel, J., & Karppi, T. (2015). Our metrics, ourselves: A hundred years of self-tracking from the weight scale to the wrist wearable. *European Journal of Cultural Studies, 18*(4–5), 479–496.

Dardot, P., & Laval, C. (2013). *The new way of the world: On neoliberal society*. London: Verso.

Day, R. E. (2014). *Indexing it all: The subject in the age of documentation, information, and data*. Cambridge, MA: MIT Press.

Dean, J. (2009). *Democracy and other neoliberal fantasies: Communicative capitalism and left politics*. Durham, NC: Duke University Press.

Deloitte. (2015). *Global human capital trends 2015: Leading in the new world of work*. London: Deloitte University Press.

Desrosières, A. (1998). *Speaking Against Number: Heidegger, Language and the Politics of Calculation*. Edinburgh: Edinburgh University Press.

Elden, S. (2006). *Speaking Against Number: Heidegger, Language and the Politics of Calculation*. Edinburgh: Edinburgh University Press.

Espeland, W. N. (1997). Authority by the numbers: Porter on quantification, discretion, and the legitimation of expertise. *Law and Social Inquiry, 22*(4), 1107–1133.

Espeland, W. (2015). Narrating numbers. In R. Rottenburg, S. E. Merry, S. J. Park, & J. Mugler (Eds.), *The world of indicators: The making of governmental knowledge through quantification* (pp. 56–75). Cambridge: Cambridge University Press.

Espeland, W. N., & Sauder, M. (2007). Rankings and reactivity: How public measures recreate social worlds. *American Journal of Sociology, 113*(1), 1–40.

Espeland, W. N., & Stevens, M. L. (2008). A sociology of quantification. *European Journal of Sociology, 49*(3), 401–436.

Featherstone, M. (2000). Archiving cultures. *British Journal of Sociology, 51*(1), 168–184.

Featherstone, M. (2006). Archive. *Theory Culture and Society, 23*(2–3), 591–596.

Foucault, M. (2002b) *Power: Essential works of Foucault 1954–1984* (Vol. 3). London: Penguin.

Foucault, M. (2007). *Security, territory, population: Lectures at the Collège de France 1977–1978*. Basingstoke: Palgrave Macmillan.

Foucault, M. (2014). *On the government of the living: Lectures at the Collège de France 1979–1980*. Basingstoke: Palgrave Macmillan.

Gane, N., & Beer, D. (2008). *New media: The key concepts*. Oxford: Berg.

Gerlitz, C., & Helmond, A. (2013). The like economy: Social buttons and the data-intensive web. *New Media and Society, 15*(8), 1348–1365.

Gerlitz, C., & Lury, C. (2014). Social media and self-evaluating assemblages: On numbers, orderings and values. *Distinktion: Scandinavian Journal of Social Theory, 15*(2), 174–188.

Graham, S., & Marvin, S. (2001). *Splintering urbanism: Networked infrastructures, technological mobilities and the urban condition*. London: Routledge.

Grosser, B. (2014). What do metrics want? How quantification prescribes social interaction of Facebook. *Computation Culture, 4*. Accessed August 14, 2015, from http://computationalculture.net/article/what_do_metrics_want

Halpern, O. (2014). *Beautiful data: A history of vision and reason since 1945*. Durham, NC: Duke University Press.

Harding, L. (2015, September 25). The node pole: Inside Facebook's Swedish hub near the artic circle. *The Guardian*. Accessed October 4, 2015, from http://www.theguardian.com/technology/2015/sep/25/facebook-datacentre-lulea-sweden-node-pole

Hayles, N. K. (1999). *How we became posthuman: Virtual bodies in cybernetics, literature, and informatics*. Chicago: The University of Chicago Press.

Hearn, A. (2008). 'Meat, mask, burden': Probing the contours of the branded 'self'. *Journal of Consumer Culture, 8*(2), 197–217.

Hearn, A. (2010). Structuring feeling: Web 2.0, online ranking and rating, and the digital 'reputation' economy. *Ephemera, 10*(3–4), 421–438.

Hill, D. W. (2015). *The pathology of communicative capitalism*. Basingstoke: Palgrave Macmillan.

Introna, L. D. (2015). Algorithms, governance, and governmentality. *Science, Technology and Human Values*. Online first. doi:10.1177/0162243915587360.

Karppi, T., & Crawford, K. (2015). Social media, financial algorithms and the hack crash. *Theory, Culture and Society*. Online first. doi: 10.1177/0263276415583139.

Kember, S., & Zylinska, J. (2012). *Life after new media: Mediation as a vital process*. Cambridge, MA: MIT Press.

Kennedy, H. (2015, July 22). Seeing data: Visualisation design should consider how we respond to statistics emotionally as well as rationally. *LSE Impact*

Blog. Accessed November 18, 2015, from http://blogs.lse.ac.uk/impactof socialsciences/2015/07/22/seeing-data-how-people-engage-with-data-visualisations/

Kennedy, H. (2016). *Post, Mine, Repeat.* Basingstoke: Palgrave Macmillan.

Kennedy, H., Hill, R., Aiello, G., & Allen, W. (2016). The work that visualisation conventions do. *Information, Communication & Society, 19*(6), 715–735.

Kennedy, H., & Moss, G. (2015). Known or knowing publics? Social media data mining and the question of public agency. *Big Data and Society*, 1–11. doi: 10.1177/2053951715611145.

Kitchin, R. (2014b). *Thinking critically about and researching algorithms.* The Programmable City. Working Paper 5, 28 October 2014.

Kitchin, R., & Dodge, M. (2011). *Code/space: Software and everyday life.* Cambridge, MA: MIT Press.

Knight, W. (2015, July 7). Inside Amazon's warehous, human-robot symbiosis. *MIT Technology Review.* Accessed November 18, 2015, from http://www.technologyreview.com/news/538601/inside-amazons-warehouse-human-robot-symbiosis/

Konings, M. (2015). *The emotional logic of capitalism: What progressives have missed.* Stanford, CA: Stanford University Press.

Lash, S. (2007). Power after hegemony: Cultural studies in mutation. *Theory Culture and Society, 24*(3), 55–78.

Law, J., & Ruppert, E. (2013). The social life of methods: Devices. *Journal Cultural Economy, 6*(3), 229–240.

Lemke, T. (2011). *Bio-politics: An advanced introduction.* New York: New York University Press.

Lovink, G. (2011). *Networks without a cause: A critique of social media.* Cambridge: Polity Press.

Lupton, D. (2013). Understanding the human machine. *IEEE Technology and Society Magazine*, Winter 2013.

Lupton, D. (2014). Apps as artefacts: Towards a critical perspective on mobile health and medical apps. *Societies, 4*, 606–622.

Mackenzie, A. (2006). *Cutting code: Software and sociality.* New York: Peter Lang.

MacKenzie, D. (2014, December 4). *At Cermak.* London Review of Books, p. 25.

Mair, M., Greiffenhagen, C., & Sharrock, W. (2015). Statistical practice: Putting society on display. *Theory, Culture and Society.* Online first. doi: 10.1177/0263276414559058.

Manovich, L. (2013). *Software takes command.* London: Bloomsbury.

Mason, P. (2015). *Postcapitalism: A guide to our future*. London: Allen Lane.

Mirowski, P. (2013). *Never let a serious crisis go to waste: How neoliberalism survived the financial meltdown*. London: Verso.

Nafus, D., & Sherman, J. (2014). This one does not go up to 11: The quantified self movement as an alternative big data practice. *International Journal of Communication, 8*, 1784–1794.

Neff, G. (2013). Why big data won't cure us. *Big Data, 1*(3), 117–123.

Osborne, T., Rose, N., & Savage, M. (2008). Reinscribing British sociology: Some critical relfections. *The Sociological Review, 56*(4), 519–534.

Parikka, J. (2010). *Insect media: An archaeology of animals and technology*. Minneapolis: University of Minnesota Press.

Parikka, J. (2012). *What is media archaeology?* Cambridge: Polity Press.

Parisi, L. (2013). *Contagious architecture: Computation, aesthetics, and space*. Cambridge, MA: MIT Press.

Pasquale, F. (2015). *The black box society: The secret algorithms that control money and information*. Cambridge, MA: Harvard University Press.

Porter, T. M. (1995). *Trust in numbers: The pursuit of objectivity in science and public life*. Princeton, NJ: Princeton University Press.

Porter, T. M. (2015). The flight of the indicator. In R. Rottenburg, S. E. Merry, S. J. Park, & J. Mugler (Eds.), *The world of indicators: The making of governmental knowledge through quantification* (pp. 34–55). Cambridge: Cambridge University Press.

Rieder, B. (2012). What is in pagerank? A historical and conceptual investigation of a recursive status index. *Computational Culture 2*. Accessed August 14, 2015, from http://computationalculture.net/article/what_is_in_pagerank

Rose, N. (1999). *Governing the soul: The shaping of the private self* (2 ed.). London: Free Association Books.

Rottenburg, R., & Merry, S. E. (2015). A world of indicators: The making of governmental knowledge through quantification. In R. Rottenburg, S. E. Merry, S. J. Park, & J. Mugler (Eds.), *The world of indicators: The making of governmental knowledge through quantification* (pp. 1–33). Cambridge: Cambridge University Press.

Ruppert, E,. & Savage, M. (2012). Transactional politics. *Sociological Review, 59*(s2), 73–92.

Sampson, T. D. (2012). *Virality: Contagion theory in the age of networks*. Minneapolis: University of Minnesota Press.

Savage, S. (2009). Against epochalism: An analysis of conceptions of change in British Sociology. *Cultural Sociology, 3*(2), 217–238.

Savage, M. (2010). *Identities and social change in Britain since 1940: The politics of method*. Oxford: Oxford University Press.

Savage, M. (2013). The 'Social life of methods': A critical introduction. *Theory Culture and Society, 30*(4), 3–21.

Savage, M., & Burrows, R. (2007). The coming crisis of empirical sociology. *Sociology, 41*(5), 885–900.

Skeggs, B., & Yuill S. (2015). Capital experimentation with person/a formation: How Facebook's monetization refigures the relationship between property, personhood and protest. *Information, Communicaton and Society*. Online first. doi: 10.1080/1369118X.2015.1111403.

Striphas, T. (2015). Algorithmic culture. *European Journal of Cultural Studies, 18*(4–5), 395–412.

Thrift, N. (2005). *Knowing capitalism*. London: Sage.

Turow, J. (2006). *Niche envy: Marketing discrimination in the digital age*. Cambridge, MA: MIT Press.

Turow, J., McGuigan, L., & Maris, E. R. (2015). Making data mining a part of life: Physical retailing, customer surveillance and the 21st century social imaginary. *European Journal of Cultural Studies, 18*(4–5), 464–478.

Urry, J. (2003). *Global complexity*. Cambridge: Polity Press.

van Doorn, N. (2014). The neoliberal subject of value: Measuring human capital in information economies. *Cultural Politics, 10*(3), 354–375.

Wang, X. (2014, October 28). From attention to citation: What are almetrics and how do they work? *LSE Impact Blog*. Accessed July 6, 2015, from http://blogs.lse.ac.uk/impactofsocialsciences/2014/10/28/from-attention-to-citation-what-and-how-do-altmetrics-work

4

Possibility

So far we have focussed upon measurement and the circulation of metrics. In this chapter the focus is drawn towards an understanding of the elusive ways in which these circulating measures come to intervene in the performance of the social world. In short, this chapter looks at how circulating metrics may have implications for what is possible. It does this on four related fronts, all of which extend themes that have been identified in the earlier chapters. First, it looks at the relations between measurement and inequality. The key observation in this section is that competition is designed to create unequal outcomes, as such metrics are at the heart of the production and maintenance of inequalities. It then moves to explore the way that judgements about value and worth shape what is seen to be possible and desirable. This section explores how metrics are used to define value and to shape notions of what is worthwhile. In the third section, the chapter then looks at the relationship between possibility and visibility. This section thinks about the power of what is seen and what is unseen. Finally, the chapter returns to some of the themes from Chap. 2 to explore the differences between probability and possibility. In this instance, the focus is upon metrics contributing to imagined futures that are then used to inform decision making.

© The Editor(s) (if applicable) and The Author(s) 2016
D. Beer, *Metric Power*, DOI 10.1057/978-1-137-55649-3_4

The overarching argument of this chapter is that the power of circulating metrics is in how they define what is possible and what is seen to be possible. The four sections in this chapter all, in different ways, illustrate how measurement relates directly to questions of possibility. This fourth chapter extends the analysis by suggesting that *metric power* cannot just be understood as a form of circulating metrics, although this is crucial, but to fully understand it, we need to also think about what these circulating metrics make possible—which is to say, what they make appear to be possible. As I will describe, metric power shapes what is possible by marking out divisions, by defining value, by rendering visible, and by envisioning outcomes.

Possibility and Inequality: What Are the Chances of That?

The relations between possibility and inequality are, of course, far-reaching and have been at the centre of much social research. Arundhati Roy's (2014) vision of the 'ghosts' of capitalism, which charts the social impact of various forms of unequal distribution and opportunity, is one such powerful and recent intervention into inequality and its human consequences. Of course, then, we have popular debates about social class and the possibilities, or lack thereof, for social mobility (for a recent example of this type of work, see the special issue on the Great British Class Survey and 'Sociologies of Class' published by *The Sociological Review* 2015; or the recent popular book by Savage 2015). And attached to this is the recent interest in the relations between culture and social class, in terms of the connections between cultural engagement and notions of class-based distinction (such as in Bennett et al. 2009) or the differential treatment of certain social classes in media content, most notably in detrimental and disproportionate visions of certain groups (Skeggs 2005) and the rise of a new 'class pantomime' (Tyler and Bennett 2010). Imogen Tyler's (2013: 4) recent and powerful work on 'social abjection' is particularly notable in terms of this type of stigmatisation of social groups. The current resurgence of interest in the super-rich explores the lives of the small minority for whom inequality creates plenty of possibilities and extremes of choice, elective segregation, and luxury (Atkinson and Burrows 2014). And then, of course, there has been the rolling debate over the merits of Piketty's (2014) detailed accounts of inequality and capital (see the special issue of the *British Journal of Sociology*, edited by Dodd, 2014). On a more ethnographic scale, we also have important accounts of what inequality means for the lived experience, particularly where marginalised lives are vilified, such as in Lisa Mckenzie's (2015) recent accounts of the lived experiences of austerity on the St Anne's estate in Nottingham. In parallel to this, and giving further context to Mckenzie's work, Jamie Peck (2012) has provided a geographical analysis of the distribution effects of austerity in the USA. If we return to the themes covered in Chap. 1, we can also add in discussions of the disproportionate impacts of neoliberalism in terms of gender (see Scharff 2014; Oksala 2013). In short, debates about the connections between possibility and inequality are still very lively today. Indeed, the conditions of global austerity seem to have re-energised these debates anew.

The above outline is by no means an exhaustive list, but it captures some recent currents and gives a feel for the broad scope of the current debates. We can see that there is seen to be much need to think in terms of the wide-ranging inequalities that are at play in the social world. In the context of this book though, we might draw our focus more directly on what measurement and the circulation of metrics might mean for these relations between possibility and inequality. To give one example, we might wonder, for instance, what role metrics might play in the 'production of abject subjects' to which Tyler (2015) refers. What part do various measures play in enabling and targeting abjections, with statistics like benefits claim amounts, medical expenses, costs to policing, welfare burdens, antisocial behaviour rates, often evoked to produce or demarcate and fuel abjection. This is a question for another time, but it might begin to reveal the underlying presence of metrics in the promotion and justification of persistent or new inequalities on a broader scale.

In terms of the role of measurement and metrics in the creation and perpetuation of inequality and unequal distributions of possibilities, Ian Hacking (1990: 6) has argued that 'enumeration requires categorization, and that defining new classes of people for the purposes of statistics has consequences for the ways in which we conceive of others and think of our own possibilities and potentialities'. Hacking (1991), as we saw in Chap. 2, uses the 'idea of making up people' in order to understand how we are created through statistics and the way in which those statistics come to be organised—in the form of categories, classification, and types. The now ubiquitous social media tag can be added to this list (see Beer 2013: 53–61). As Hacking (1991: 194) claims, the 'bureaucracy of statistics imposes not just by creating administrative rulings but by determining classifications within which people must think of themselves and of the actions that are open to them'. So in this formulation, measurements *make us up*, and in so doing, define the lines of inequality, thus shaping what is possible, what opportunities are presented, and what the limits on our lives might be. We should make clear that these are not simply accepted but that all kinds of important 'classificatory struggles' (Tyler 2015) unfold, leaving us with little doubt that classificatory boundaries are defining of inequalities. Thus, Imogen Tyler (2015: 507) discerns that the way to respond is not to pursue classifications in the analysis of the social world but to aim at 'exposing and critiquing the consequences of classificatory systems and the forms of value, judgments and norms they establish in human societies'. Any attempt to understand the power of metrics is likely to follow Tyler's broader sentiment here—the challenge is to understand the metrics and the classifications that order them as they constitute and afford notions of difference and value.

For Hacking, as for others, it is the classification of population statistics that is particularly powerful in these processes and in the consequences of the measures about us. The world needed to be regarded or thought of as being measurable in order for this to occur (see Chap. 2 for a discussion of this). Hacking (1990: 5) argues, for instance, that the 'imperialism of probabilities could occur only as the world itself became numerical'. The spread of probabilistic calculations about the world could only happen once that world is seen to be measurable and categorisable (and as ways of measuring are expanded). These conditions have enabled the spread of calculatory and probabilistic approaches to understanding the social and natural world.

The consequences of the classification of the metrics about us escalate in significance as the data accumulate at an accelerating pace. Referring to an earlier era, Hacking's observation is that:

[t]he avalanche of numbers, the erosion of determinism, and the invention of normalcy are embedded in the grander topics of the Industrial Revolution. The acquisition of numbers by the populace, and the professional lust for precision in measurement, were driven by familiar themes of manufacturing, mining, trade, health, railways, war, empire. Similarly the idea of a norm became codified in these domains. (Hacking 1990: 5)

Here we see how an increasing accumulation of numbers might also mean an increasing cementing of norms and notions of normalcy. Hacking takes us back to the Industrial Revolution to suggest that the 'avalanche of numbers' was a consequence of the pursuit of precise measurement across a range of social sectors. The presence of these numbers made it possible to find norms and for these norms to then become realities of sorts (for a more recent version of this argument, see Day 2014: 135). We can imagine then that measurement is powerful not just in creating norms but in envisioning them and enabling them to circulate into the social world in different forms. Beyond the power of classification then, we might also see the formulation of norms in these numbers as being powerful in shaping and maintaining social divisions of different sorts. More powerful systems of measurement are likely to lead, via a faith in those numbers, to more obstinate and obdurate notions of normalcy.

When considering the possibilities that measures afford, it is then crucial to think also of the part played by categories and classifications. These categories are powerful. As Espeland (1997: 1117) explains, 'when statistical categories are bolstered by the authority of powerful institutions, however artificial or superficial they appear, they become real and durable'. These categories, irrelevant of their failings, become solid and obdurate. With the emergence of numerical thinking and ways of measuring, categories emerged that grouped people and contributed to the establishment of norms. But these were often existing categories that were imposed upon systems of counting, thus reinforcing existing or enabling reconfigured and re-engineered divisions and notions of difference. Hacking (1990: 3) argues that:

[c]ategories had to be invented into which people could conveniently fall in order to be counted. The systematic collection of data about people has

affected not only the ways in which we conceive of a society, but also the ways in which we described our neighbour. It has profoundly transformed what we choose to do, who we try to be and what we think of ourselves.

In this account, the result of counting and categorisation is both deeply social and personal. These categories that are used to count people shape how society is understood, conceived, comprehended, and approached. These come to play a part in how we understand people, in choices and behaviours, and even in our own senses of identity. Hacking's arguments here are compelling, and show just how powerful metrics might be in drawing the possibilities of the social world.

Underpinning this is the role that numbers and numerical categories play in defining what is seen to be normal, appropriate, and justifiable. These numbers create robust social laws that regulate and limit individual choice and social conduct, and which then contribute to how people are judged or perceived. According to Hacking (1990: 2):

> [s]uch social and personal laws were to be a matter of probabilities, of chances. Statistical in nature, these laws were nonetheless inexorable; they could even be self-regulating. People are normal if they conform to the central tendency of such laws, while those at the extremes are pathological. Few of us fancy being pathological, so 'most of us' try to make ourselves normal, which in turn affects what is normal. Atoms have no such inclinations. The human sciences display a feedback effect not to be found in physics.

The numbers provide us with accounts of normalcy that we mostly try to adhere to. The result is that such norms are cemented as new measurements of the social world take place. Guided by the numbers, we make ourselves normal, which then reinforces normalcy. Thus we find a kind of recursive establishment of normalcy as circulating calculative norms bury themselves into the flesh of the social world. Measuring the social world is different from measuring the natural world, Hacking concludes in the above passage. When measuring the social world there are inevitable feedback loops, with the result being that behaviours adapt in response to the measures. The social world has feedback loops that the natural world does not. Thus the numbers that we use and the categories they are counted within set laws, establish norms, and become a part of the social world. We are returned here to the arguments about the 'social life of the methods' (Savage 2013) discussed in Chap. 3. Here, in Hackings writings, we find a profound version of methods becoming part of the functioning of the social world. The way we are counted and the way that we are categorised inevitably becomes part of that social world that is being measured. Hacking's use of feedback loops is important here. It gives the sense that we are pursuing an ever more cemented and concrete version of normalcy, with the feedback loops tightening and strengthening the perception of the acceptable from the different, the normal from the abnormal, and so on.

This discussion of the power of norms and categories, of course, brings us to the question of choice. Gordon (1991: 43) observes that neoliberalism uses the notion of 'choice' as a powerful means of promoting its agenda, this notion of people having choice and the emphasis on the importance of choice in freedom 'empowers economic calculation to sweep aside the anthropological categories and frameworks of the human and social sciences'. A neoliberal approach is to suggest that choice somehow breaks down the types of social laws to which Hacking has referred. But this is only by replacing one kind of calculation for another. We can return here to some of the issues with which we opened this book, and particularly the notion of competition as an organising principle in the social world. In that opening chapter, we discussed how the neoliberal art of government, to use Foucault's terminology, is based upon the implementation of forms of competition, or at least the implementation of *the mechanisms of competition* as I referred to it. In Chap. 1, I argued that we should see metrics as being the mechanisms by which competition is realised. Competition, clearly, is based upon the production of inequalities of different sorts. As such, metrics are the tool by which these inequalities are created and maintained, which of course shapes the opportunities and possibilities for those who are measured and their subsequent placing in the field of competition. Returning again to Will Davies' key insights on competition, he argues that a 'society that celebrates and encourages "competitiveness" as an ethos, be it in sport, business, politics or education, cannot then be surprised if outcomes are then highly unequal' (Davies 2014: 37). As I discussed in Chap. 1, it is argued that 'reconfiguring institutions to *resemble* markets is a hallmark of neoliberal government' (Davies 2014: 38, italics in the original). The consequences of this are clear, as competition spreads and expands so do the particular inequalities that it affords and realises. Neoliberalism, we have seen to be argued, 'depends precisely on *constructing or imputing certain common institutional or psychological traits, as preconditions of the competitive process*' (Davies 2014: 37, italics in the original). These common traits promote competitiveness whilst also giving the common grounds upon which competition can be exercised. This does not mean that outcomes are seen to be inflexible, indeed the very notion that competition is open and fair, with clear defining rules, takes on an important role in the rhetoric. This rhetoric makes competition an appealing thing despite the potential for the unequal outcomes that it creates. As Davies (2014: 37) puts it, the 'great appeal of competition, from the neoliberal perspective is that it enables activity to be rationalized and quantified, but in ways that purport to maintain uncertainty of outcome' (I will discuss this production of uncertainty in much more detail in Chap. 6). Competition is appealing then because it facilitates the quantified rankings and differentiations that, from a neoliberal perspective, are seen to be favourable and productive. The metrics of competition appear to make outcomes that are rational, logical, and objective (as we have already seen in Chap. 2), even if they are not. The important point for Wendy Brown (2015b: 42) is that 'inclusion inverts into competition' in a neoliberal rationality—thus it 'intensifies inequalities'. You are *included* in these competitions and in the gathering of metrics so that you can be *judged* and *ranked*.

In his exploration of neoliberalism's limits, Will Davies refers to the 'paradoxes of competition'. The first of these paradoxes concerns the role of the state. The state is both 'active' and 'disengaged' (Davies 2014: 40). This is the question of intervention that is discussed in detail in Foucault's (2008) lectures on *The Birth of Biopolitics* (and which was discussed briefly in Chap. 1). The interventions come, in a neoliberal mode of governance, only to ensure the mechanisms for competition operate. As such the state is both active and disengaged. The second paradox though is where we begin to see what this means for inequality. As we have noted, competition is designed to produce unequal outcomes—with metrics being the means by which this inequality is calculated and differentiated. The second paradox that Davies examines is competition's paradoxical combination of 'equality and inequality'. As Davies (2014: 54) explains, at least a 'sense of equivalence' is needed in the first instance, from which then the differential outcomes might be communicated. This is the 'tension between "equality" and "inequality" [that] sits at the heart of any competitive event or activity' (Davies 2014: 57). A shared sense of equivalence is needed to give some grounds on which to compare competitive entities, but these are then used as the basis to cultivate a sense of fairness whilst producing unequal and differential outcomes, rankings, and judgements. Thus, a sense or notion of equality and fairness can be used to justify the production and maintenance of inequality.

This is why, referring to transitions in the seventeenth century, Foucault has argued that 'statistics…now becomes the main technical factor, or one of the main technical factors, in unblocking the art of government' (Foucault 2007: 104). This way of thinking was already present back then, but it was blocked by the limited technological infrastructures of the time. If we extend Foucault's point we might conclude here that technological changes and the rise of metrics have enabled the continued unblocking of this mode of governance and allowed it to continue to expand and even flourish. The new statistics, of which we might include the recent interest in big data, has the capacity to continue to unblock such a set of interests.

In short then, competitors need to have the sense that there is some shared and equal ground upon which to compete, which in turn is what produces and legitimises unequal outcomes. The inequalities produced through competition and the metrics that afford that competition shape what is possible and what chances are available. Measures circulate through the social world ensuring inequality through competition and thus shaping life-chances and opportunities. This gives us something of a broad starting point, but forces us to ask within this more general framework in what ways circulating metrics might shape what is possible, what is unlikely, and what is impossible. The questions this raises, both in this chapter and in Chap. 3, concern both value and visibility.

Value, Worth, and Possibility: What Is Valued? What Is Worthwhile? What, Then, Is Possible?

Value and worth are, of course, slippery concepts. Metrics are frequently used though to give some apparent if superficial solidity to them. With metric forms of knowledge we can see how notions of value and worth become something of a battleground in the shaping of what is possible. What is of value or worth is likely to be what is encouraged or endorsed by metrics and, therefore, is likely deemed the route to be taken. Ronald Day turns to the type of discourse around big data to reflect on such questions:

> The data says…; the data shows us…; we are only interested in data [not justi-
> fications/excuses/your opinion/your experience]…; big data and its mining and
> visualizations gives us a macroscopic view to see the world anew now—these
> and similar phrases and tropes now fill the air with what is claimed to be a
> new form of knowledge and a new tool for governance that are superior to
> all others, past and present. (Day 2014: 134; italics in the original)

As he puts it, these 'claims for knowledge are presented as imme-diate—"factual"' (Day 2014: 134). The concept of 'big data' becomes the means by which certain viewpoints or conclusions are justified and promoted. In these formulations, the presence or recourse to data is the means by which avenues are closed and opinions or responses are jetti-soned as value is demarcated and possibilities are defined. The discourse of data is as powerful in defining what is possible as the data itself, if not more so. What we can take from Day's claims is that numbers are hard to argue against—they are convincing; they leave little ground for any subjective response or reaction. Data is simply seen to be better than anything any human intuition or judgement might offer. The data says this—that is the end of the story. Part of the power of metrics is to be found in the way that these metrics are spoken about and regarded—with the data seen to be powerful whilst human agency is seen as poten-tially unreliable, inefficient, and limited in the depth of its analytic gaze and impartiality.

Elsewhere, the power of discourse around big data is also seen to shape perceptions of their potency. In his recent book, Rob Kitchin (2014a: 126) has claimed that:

> [t]he power of the discursive regimes being constructed is illustrated by considering the counter-argument—it is difficult to contend that being less insightful and wise, productive, competitive, efficient, effective, sustainable, secure, safe, and so on, is a desirable situation. If big data provide all of these benefits, the regime contends that it makes little sense not to pursue the development of big data systems.

This is an interesting point about the use of big data within power dynamics. The discursive regimes around big data have become hard to resist. The result is that measurement is ushered into more and more aspects of our lives, with little room for resistance or response. Big data is presented as being based around common sense. From such a perspective, big data is equated to progress. Big data is efficient and facilitates competitiveness. The arguments about being measured are being won by the very concept of big data. Automatically, these measures are seen to be valuable and to reliably uncover value. Kitchin's (2014a: 126) answer is to propose that 'what is presently required, through specific case studies is a much more detailed mapping out and deconstruction of the unfolding discursive regimes being constructed'. But those discursive regimes are likely to be fairly solid in their form, and may even have genealogical roots that are hard to shift (as was suggested in Chap. 2). We are already thinking in numerical terms; we have a calculatory mindset. Our starting point then, for thinking about how possibility and value are related, is to think about how big data are seen to have innate and unquestionable value, and that are likely to be used to locate and compare the things that are seen to be worthwhile.

This again drags us back to notions of objectivity and its construction. Such big data-based judgements about value are formed through a sense that 'quantification is a technology of distance' (Porter 1995: ix). Numbers appear to place their users at a critical and objective distance from the object that is being judged and the decision that is then being made. If calculation provides a sense of distance, then quantification is providing the opportunity to judge value at a distance and to make such judgements appear hyper-rational, fair, and indisputably logical. As such, the very notion of value or worth becomes one of calculation. Calculation provides the means by which value can be governed. There are two things to consider here. First, measures demarcate value. Second, the measures then become the means by which those values and preferred outcomes can be pursued. As Porter (1995: 45) succinctly explains:

> the measures succeed by giving direction to the very activities that are being measured. In this way individuals are made governable; they display what Foucault called governmentality. Numbers create and can be compared with norms, which are among the gentlest and yet most persuasive forms of power in modern democracies.

Measures direct activities, often in subtle ways. Notions of value and worth, I would suggest, are the means by which measures are able to direct action, behaviours, and practices (see also Chap. 6). This is an important shift that allows us to see that it is through a sense of what is worthwhile that metrics are able to shape what is possible. We are discouraged from spending time, energy, and resources on things that are not deemed worthwhile by the metrics. This would suggest that as the scale and intensity of metrics increases it will also mean that what is seen to be of value will increasingly be demarcated for us by these systems. Thus the parameters of possibility, through the increasing measurement of value, will be set out in various forms and with the obduracy that comes with a logic of objective distance. If we listen to the metrics, we will see that they are not just capturing, they are also, often, instructing. Indeed, as Moor and Lury (2011) have described, processes of valuation are central to the way in which organisations measure the value of their brands and their brand values. These valuations, they find, then become performative and instructive.

Clearly, what is seen to be of value or worth shapes judgements and is also likely to shape behaviours by emphasising certain actions, products, or outcomes. That which is seen to lack value or to not be worthwhile is likely to be seen to be detrimental, wasteful, or harmful. This has already been hinted at a little, but it is worth further attention given the way that measurement functions to define what is valued and what is seen to be worthwhile, which in turn then shapes what is seen to be desirable and therefore possible.

A key development in this pursuit of notions of value is what Higgins and Larner (2010a, b) define as 'standards'. It is through standards that the 'calculation of the social' can occur, it is through standards that measures can be compared or where they come into some sort of analytical unison. As we have seen through the work of Will Davies (2014: see also Chap. 1 and also this chapter), for competition to function, some key similarities and standards are needed—shared standards provide one such means of competitive comparison and shared grounds for competition. These standards, Higgins and Larner (2010a: 3) propose, form into 'assemblages of regulation'. They warn that we should resist seeing these standards as universal or fixed but, nevertheless, we might see them as producing barriers that limit possibilities of different sorts. Here, standards are also to be understood as 'objects of knowledge' (Higgins and Larner 2010b: 205) that can be reworked but which provide powerful obstacles and parameters within which these calculations of the social operate.

It is for these reasons, based on the role of metrics in the envisioning of standards, that Day talks of the relations between potentiality and possibility. For Day (2014: 136; italics in the original), *governance using documentary systems must turn the potential into the possible, and so fit the person within logical systems of representation*. As such, these systems draw us into what is seen to be the logical frames of dominant representations. Those powerful norms that we have spoken of, that are cemented by metrics, in turn limit *potential* by turning it into the specifics and bounded limits of *possibility*. This turning of the openness of potential into the limits of representational accounts of possibility is a product of the possibilities of the systems that order and facilitate the circulation of data (as discussed in Chap. 3). We can also see this type of transformation in other terms. For instance, Doria (2013: 19) has written of the 'relationship between calculability and mobilization of the self'. In this formulation, the self is mobilised and activated in response to the calculations to which it is exposed. This is something which has already been discussed, but in this instance, we can see how the shift from the potential to the possible is translated into self-actualisation. Once it is exposed to calculation, the self shifts from being a site of potential to a site of limited possibilities. We can see here how we are returned to the self-training and self-disciplining neoliberal subject discussed in Chap. 1, and we begin to see directly how this self-training is a product of the possibilities and guidance that are put in place by these systems.

If we bring these observations back to notions of value and worth, then we can begin to see how a sense of value is likely to be caught up in the logic and standards of the system. Again, possibility is defined by notions of value as we move from potential to possibility. Bev Skeggs (2014) has recently articulated this in terms of the relations and tensions between value and values. The question she poses is about the way in which those things that are valued may not fit with our values, indeed these things can be in tension with one another. Skeggs wonders if there is a space to defend or reassert our values once they are overturned and usurped by the pursuit and logic of economic value. Working with a slightly different terminology but with an interest in a similar set of concerns, Doria (2013: 43) points out that 'the calculable and the incalculable are considered to be living in a relationship of co-othering, precisely on the basis of their common nature as enacted entities'. The incalculable here is complex. There is a sense in Skeggs' work that we need to try to defend the incalculable from the logic of capitalism and its growing cultures of measurement. Two issues arise here. The first is that it is possible that proxy forms of data can be co-opted to mean that what is considered incalculable may be rendered calculable at any point (if so desired), particularly where the infrastructures of data harvesting include many aspects of bodily movements and routines. So the inescapability of metrics needs to be thought of in terms of the complexity of the use of proxy measures. Second, those things that lie outside of calculation, and for which there is no desire to locate a proxy measure, will be those that are not considered to be of value. As Badiou (2008: 2) has recently reiterated, 'what counts—in the sense of what is valued—is that which is counted' (see also White 2014: 131). Thus, this 'privileges the calculable over the incalculable', Venn (2009: 226) explains, and also 'reduces the incalculable to the status of the calculable'.

Taken together, the relations between values and value or between the calculable and the incalculable fall into what Doria (2013: 168) has called 'the problem of measure' (see also Halpern 2014: 25). The problem here is that ideas around quality, as an assessment of value, are limited to a 'calculative reduction' (Doria 2013: 153)—Lisa Adkins (2009), for instance, talks of the 'crisis of measure' that can be associated with immaterial labour and its impact upon the lives of women, particularly relating to time and temporality. But this need not be limited to workplace performance. As we have already discussed, with devices like smartphones or wearables like FitBit or the Apple Watch we begin to see extremes of how metrics might be used to monitor, shape, and hone the performance of our social life and bodily routines. As I discussed at the opening of the book, the well trained, active, and engaged self-monitoring individual would appear to be the aim of such devices. Here notions of quality and value are transposed directly onto the body and the lifestyle choices of individuals. And of course trackable devices are already an established part of many people's working lives. As a further example, we have also already noted the performance training properties of social media (see Chap. 3).

Now of course this is not to say that there is a passive acceptance of the values that are implicit in metrics, nor of the possibilities that these project. In fact, there is a need to greatly expand our understanding of how people might defend the incalculable, protect values, or find ways of reinterpreting the possibilities defined by the metrics to which they are exposed. In his wide-ranging book on the escalation of data, for instance, Rob Kitchin (2014a: 127) notes that 'people start to game the system in rational, self-interested but often unpredictable ways to subvert metrics, algorithms, and automated decision-making processes'. We need to keep this important observation in mind so as not to take a reductive approach to metrics that in some way suggests that we are all sedentary and subservient in their presence—and we are reminded again about Espeland and Sauder's (2007: 29) use of the concept of 'reactivity' and also how they apply this to 'gaming rankings'. Despite this though, we can see in Kitchin's point that such responses often operate within the logic of these systems—people find ways to play the system so that they do well in the metrics. This then shows how metric power works by producing certain behaviours or outcomes (Beer 2015b)—often these preferred outcomes will be intended but on other occasions they will be a product of people living with metrics and then learning ways to subvert and contort their practices to suit the measures (and any loopholes that might exist in these systems). To go back to the example of the call centre with which I opened Chap. 1, hanging up on customers to quickly take the next call was a good way to score a high call rate. Although, to add, this was a practice that could be spotted easily if the average call length was too short. So even gaming the system is to work within or be judged within its logic (I will return to this issue in Chap. 6).

When thinking about how people are positioned by the judgements of value that are implicit in these systems, we can reflect upon the variable possibilities that people have for working the metrics to their advantage. Some are better placed than others to find their worth or to have calculable value. The possibilities that are associated with the measurement of value are likely to be far from evenly distributed. When thinking of the uneven distribution of the possibilities that come with the measurement of value we can reflect upon, to pick just one telling instance, the new industry of reputation management. This industry is designed to enable those with enough money to manage the circulation of metrics and other data that concerns them—and thus to manage their place in what Hearn (2010) refers to as the digital 'reputation economy'. These companies, such as reputation.com, provide services for managing the content and ratings that appear in social media and thus manage the social media reputation of the brand, organisation, or even the individual. These services are particularly focussed on managing any negative impressions or anything that might reduce the reputational sense of value. This means that some people and organisations are better placed to manipulate the metrics and to game the system with skilled assistance. Pasquale's (2015: 55) discussion of such an industry leads him to conclude that only those with significant economic capital are able to 'develop foolproof versions of their own personal black boxes'. Clearly, such reputation management is an indicator that the possibilities of circulating metrics are not as fixed for some as they are for others. We cannot then expect them to be value free or to circulate in evenly distributed ways. Metric power is not about the rigid application of metrics, but about the manipulation of those circulations to suit the desired possible outcomes of those with sufficient economic sway.

Indeed, for Pasquale, there is a kind of inescapability and inevitability about the way we are likely to be treated by these systems—unless we have the power to change those rules or to reverse engineer their outcomes. However good we might be at playing the system, we are still judged by them. It is out of our hands and open to the types of divisions that we find in broader social constellations of difference, stigma, and prejudice. According to Pasquale (2015: 35):

> [a]utomated systems claim to rate all individuals in the same way, thus averting discrimination…But software engineers construct the datasets mined by scoring systems; they define the parameters of data-mining analyses; they create the clusters, links, and decisions trees applied; they generate the predictive models applied. Human biases and values are embedded into each and every step of development. Computerization may simply drive discrimination upstream.

We are back then to the values that are implicit in these systems. Thus, metrics can be discriminating in *posterior* and a priori ways—they have actions both before and after we are measured (I'll discuss this further in Chap. 6). First, we have value judgements built into the systems, laced into the algorithms, archives, and the data they draw upon (as discussed in Chap. 3). Second, we have those who are empowered to do so finding ways of manipulating and reshaping the outcomes of these systems, particularly where they fall out of their direct control on decentralised platforms like social media (in the form of ratings, reviews, likes, favourites, followers, and other rankings). Social divisions then are woven into metrics and are also a material part of how they are received, both of which then dictate what is possible and what is seen to be possible.

It is then perhaps an obvious statement to say that metrics clearly are not as objective as they are often presented in the rhetoric (as we discussed in relation to measurement in Chap. 2). This becomes even clearer when we reflect upon the circulation of those metrics and the possibilities they then afford. Pasquale (2015: 61) reiterates this issue, which was explored in Chap. 2, by arguing that 'despite their claims to objectivity and neutrality, they are constantly making value-laden, controversial decisions'. The important point here is that they are making what Pasquale refers to as 'value-laden' decisions. The very arrival at such decisions seeks to perpetuate and enhance certain value positions. This leads Pasqaule (2015: 218) to conclude that 'black box services are often wondrous to behold, but our black box society has become dangerously unstable, unfair, and unproductive'. Pasquale's point here is contentious, and would need a good deal of work to explore, but his provocation is interesting in its own right and could be used to stimulate a good deal of thinking about the way that the use of metrics is promoted as the means by which we can obtain increased fairness and productivity. Pasquale's contention is that metrics and data systems might actually be having the opposite effect—this is something we might wish to recall as we consider the opportunities we might have for challenging the logic of metrics (which is discussed in both Chaps. 5 and 6).

Before closing these reflections on the role of the relations between value, worth, and possibility under the conditions of metric power, we can briefly reflect on what this might mean for individual lives, particularly with regard to how lives are shaped in response to the demands placed upon them. In other words, this is to consider the desired properties that these systems project and how people might respond. We saw in Chap. 1 how neoliberal governance focusses upon the individual and upon shaping their behaviours through marketised competition, but then we can extend the analysis to begin to see what form and actions the individual might take under such pressures (this is discussed briefly here but is to be elaborated in more detail in relation to the affects of measures in Chap. 6).

Zygmunt Bauman's (2007) vision of a 'confessional society' is now quite familiar, with the individual feeling obligated to broadcast their private lives in the public domain via social media. His argument was that when using social media we are competing for attention, so we market ourselves as appealing commodities. I also discussed Jodi Dean and Philip Mirowski's accounts of social media in the previous chapter. Day has extended some of these visions in relation to the value placed upon aspects of individual lives by these media forms and the measurements of being a successful individual that they bring, with likes, followers, views, friend numbers, and so on (I am talking here, of course, about highly circumscribed notions of success that I am not inclined to agree with). Day's (2014: 124) contention is that:

> it is common to abstract one's self in order to put one's self 'out there' on the Internet. One creates an 'online identity' toward creating a 'brand' for one's self, which may be exchanged for some other commodities (including other persons as such), as the logic of markets permeates through all human relationships—love, marriage, labor, etc.

The process here is of abstracting oneself as an individual entity within these media forms. The very act of creating an individual profile is to create a document about our lives which contains narrative content but which also produces and displays metrics about us and about our social connections. The work that is put into a social media profile creates content, but also makes numerical tracers and markers of status, standing, and influence. Day (2014: 127) adds that the 'commodity form through which the subject enters the market place is not just through his or her "immaterial" labor, but through the appearance of one's self as unified semantic forms (i.e., as documents), within marketplaces'. Individual social media profiles become documents of individuals' lives that are opened up to market conditions, to be judged for their value on a constant basis by other users (for a discussion of the marketisation of social media profiles, see Skeggs and Yuill 2015). The metrics enable those market-based judgements to occur. This is what Hearn (2008) describes as the 'branded-self' operating social media. The individual is encouraged to find and display, or locate activities that will increase the numbers on display, and thus increase their perceived value and worth. Day (2014: 127) describes this in these terms, 'one presents images of one's self through social networks, one's romantic past is ranked and chatted about in social networks, one's recommendations are seen online, one's friends are known, one's life is valued through credit histories and the like'. The point is that very little escapes the logic of these systems, with very little being considered uncaptured or immeasurable. Everything can be quantified from relationships to friendships, through to tastes, travels, levels of activity, and consumption practices. As the above passage from Day's book would hint, these lives are being designed to suit the metric- and narrative-based markets into which they are being exposed. Individual lives then become their projections in different media and metric forms, captured and judged for their value—and designed by those individuals to suit their interpretation of the market decisions and rankings to which they will be routinely exposed (from the employment market to the rankings of status and worth).

This brings us back to the question of the measurement of value or worth. Which in turn, brings us to the role of measurement in affording competition. It has been suggested that the very notion of value has taken on greater significance as new forms of measurement have spread through the social world (Adkins and Lury 2012). Value becomes something of a watchword, a focus for the organisation of the social. Value becomes the conceptual fulcrum of the intersections between measurement and ranking. Once there are innumerable ongoing attempts to find value by making it measurable, so increasingly it becomes a preoccupation and a central motif that fuels social ordering. So, as Adkins and Lury (2012: 22) indicate in their wide-ranging introduction to their book on the relation between measurement and value:

> in post- or more-than-representational spaces, experiments in measurement and value are helping to bring into existence an expansion of the social in terms of an apparent omnipresence of value that is linked to changing relations between the quantitative and the qualitative, the extensive and the intensive, representativeness and partiality.

What they call 'experiments' in measurement and value, which is found in the pursuit of new ways of measuring value, the social is transformed in particular ways (for a discussion though of the problems of the labels in the quantitative/qualitative divide in this context, see Mair et al. 2015). The social expands into the private, emotional, and corporeal aspects of our lives (see e.g. Konings 2015; Davies 2015a). Very little escapes, and virtually everything and everyone gets drawn into the measurement of their value. They continue by posing the question that this raises for sociological work:

the expansion of the social also raises questions about how to assess the validity, adequacy and efficacy of measurement in such spaces. It is thus both because of the opportunities and dangers that such developments pose that sociology must continue to put questions of measurement and value, of quantity and quality, of subjectivity and objectivity, at the heart of the discipline. (Adkins and Lury 2012: 22)

Certainly, these types of questions should be at the heart of the discipline, but not solely in thinking about how sociology is to be conducted, but also as we reflect upon our objects of study and how they might be the sites of value-based metrics and judgements. If we are to take up Adkins and Lury's suggestions, then we will certainly need to reflect on how measurement defines value, which in turn then defines what is possible. This might represent a conceptual point from which to develop such ideas, but we would need further resources to do so. In exploring these relations, we would need to go back to the types of arguments outlined in Chap. 2, particularly in relation to the ideas around the ongoing histories of the faith in numbers and the pursuit of calculative objectivity as an organising principle of metric-based cultural interests. In thinking about questions of value and what might be seen to be worthwhile, there is a need to think about how values are defined by what is visible. That is to focus upon the importance of the visibilities that relate to the type of metric-based assessments of value that achieve prominence. This leads us to the question of visibility.

Possibility and Visibility: What Is Seen and What Is Done

As with value, visibility is quite a difficult and awkward concept. In Chap. 3, we began to reflect upon how visibility might be a product of the infrastructures of which we are a part. We reflected on the role of the archiving structures and algorithmic decision typical of contemporary media in rendering things visible (and for a more detailed account, see Beer 2013). Now though we can briefly pause to reflect on how this type of visibility also has the effect of producing possibilities. We can start with some very simple observations to orientate these discussions. Twitter makes us a recommendation of people we might like to follow; these people suddenly become visible to us, and the chances of them becoming part of our network increase. This is something similar to an automated version of what Nick Crossley (2009: 41) refers to, in his social network analysis, as the 'Granovetter effect'—with new connections between people increasing in likelihood as networks intersect. Similarly, recommendations of various sorts also perform the same task, they make music, films, TV shows, books, blog posts, news stories, and so on, visible to us thus mapping out likely possible outcomes. When we think across the type of big data revolution (Kitchin 2014a) that we have heard about and scale this up to include share dealing, algorithmic border crossing security decisions, insurance price setting, decisions about which customers to prioritise, we can then begin to see how infrastructurally defined visibility might lead to all sorts of possibilities—with metrics then driving what is seen and therefore what is likely to happen. When thinking about the possibilities that are shaped and afforded by metric power, visibility is a key component. The metrics we see or the metrics that are used by these systems to decide what we encounter—the metrics we see or the metrics that see us—are likely to produce and define possibilities. The more visible particular metrics become, the more powerful they are likely to be in shaping decisions and outcomes—other less visible metrics are not. Again here we can see how the pathways of the circulation of metrics leads to visibility, which in turn then crafts and recrafts possibilities. The power of metrics is in what is seen and what is not seen. Metric power is directed in two ways here. We have the visibility of the metrics themselves, the numbers that we encounter. Then we also have the metrics guiding various filtering mechanisms and shaping what is regarded to be of importance or of interest and selecting what should be visible.

Then, of course, we have the increasing visibility of people's lives on social media, which in turn can translate into metrics about them (or that can be used to target them through advertising or news feeds). As Pasquale (2015: 19) has put it, 'tell us everything, Big Data croons. Don't be shy. The more you tell us, the more we can help you'. We are encouraged to be visible, to narrate, and to be counted. As we have already seen in this chapter and in Chap. 3, there is an imperative to be visible placed upon us by contemporary media forms.

Taking all of these things together, Badiou's work can be used to think through some broader cultural shifts here. In one crucial passage he argues that:

> [n]umber governs cultural representations. Of course, there is television, viewing figures, advertising. But that's not the most important thing. It is in its very essence that the cultural fabric is woven by number alone. A 'cultural fact' is a numerical fact. And, conversely, whatever produces number can be culturally located; that which has no number will have no time either. Art, which deals with number only in so far as there is a *thinking* of number, is a culturally unpronounceable word. (Badiou 2008: 3)

And when we put together a number of the things covered in this book, we can perhaps begin to see what Badiou is hinting at here. We have a culture of metrics and also a cultural metrics—there is both *a cultural interest in numbers and culture that is shaped and populated with numbers.* Badiou couldn't have been referring to this directly at the time of writing, but we can certainly see how culture can be understood and represented through numbers (see e.g. Beer 2015b). Cultural representations and the way that culture is received, decoded, and understood are often directly or indirectly through metrics—from data about viewing figures, to number of followers, to number of views, to performance statistics of footballers or Rugby players, financial performance of the music industry, to who is the best performing label, streaming site, on-demand service, mobile market share holder, and so on. Badiou's point forces us to reflect on the role of number in terms of the governance of cultural representations and the visibility of those representations.

Badiou's use of governance in understanding numbers and cultural visibility draws us to reflect again on the role of calculation in Foucault's work. Some of the points raised above echo back into Foucault's lectures on governance, particularly with regard to the way that calculation makes things visible, especially where this enables the envisioning of aspects of society that may not previously have been visible, such as mortality or crime rates and the like. Foucault (2007: 79), for example, claimed that:

[a] constant interplay between techniques of power and their object gradually carves out in reality, as a field of reality, population and its specific phenomena. A whole series of objects were made visible for possible forms of knowledge on the basis of the constitution of the population as the correlate of techniques of power. In turn, because these forms of knowledge constantly carve out new objects, the population could be formed, continue, and remain as the privileged correlate of modern mechanisms of power.

Foucault is working here with some formulations that are in the stages of development that we might expect from a lecture course rather than a completed project, but the suggestion here is important and also places our discussion of the politics of visibility within a genealogical framework. Visibility has been important to power formations since the rise of statistics and the archiving and calculation of populations (see Chap. 2). The human sciences, Foucault added, 'should be understood on the basis of the emergence of population as the correlate of power and the object of knowledge' (Foucault 2007: 79). In Foucault's account, the population became visible—it formed—as a consequence of being counted. The result is that these populations could then be governed based upon the objects or features that were to be found within those calculations. In short, the history of using statistics to capture people's lives and aggregate level populations has been about making things visible so that they can then be governed, regulated, controlled, and so on. Visibility then is not suddenly of importance, but is again a product of the historical forces that we discussed in Chap. 2. Again, though, the intensification of metrics is linked to increased visibility and increased opportunities for power dynamics to be realised. Foucault is referring to basic population metrics, but the types of complex metrics we have today take this type of governance through visibility to new levels of possibility—which is something that Foucault appears to have seen coming on the horizon.

In Foucault's accounts, the results are complex, but these systems of measurement of calculation of individuals and populations are depicted as a kind of 'physics of power' (Foucault 2007: 49), which govern and restrict. These are technologies of power that are based upon visibility. For Foucault (2007: 48), 'this is to say that…freedom, both ideology and technique of government, should in fact be understood within the mutations and transformations of the technologies of power'. As the technologies of power—based upon calculation—transform and mutate so too do the possibilities that they afford. The way that freedom is understood as well as the mechanisms by which it is enforced are both reshaped as we find new ways to measure people. Similarly, Miller and Rose have more recently argued that this type of government based on calculation is also about visibility. As they explain, in these circumstances government:

> works by installing what one might term a calculative technology in the heart of the 'private' sphere, producing new ways of rendering economic activity into thought, conferring new visibilities upon the components of profit and loss, embedding new methods of calculation and hence linking private decisions and public objectives in a new way—through the medium of knowledge. (Miller and Rose 2008: 67)

Here we see how new methods and new ways of rendering things calculable are powerful in terms of their ability to make things visible. The achievement of visibility through a growing calculative technological assemblage, along with the expansion of economically informed reason, is of undoubtable significance in understanding the power of calculation. Metrics render things visible in order for them to fall into the reach of those who are in positions to manage, control, and intervene.

So, this is not just a technical rendering of visibility through technologies or the apparatus of measurement, this is also the interwoven ideological changes and cultural understandings that alter as people are measured in new ways. This is an important point for understanding metric power. If we take Facebook as a brief example, we can see that technologies for capturing lives altered with the social media profile, but it also needed changes in notions of privacy for it to work. The Facebook infrastructure archives individual lives in new ways, but the ideological and cultural shifts around the imperative to broadcast lives through Facebook profiles was a necessity for its success. Here we can see how Facebook, as an archival technology, changed how people understood and approached the very doing of everyday life. As technologies such as these media platforms mutate, Foucault expects also that they reshape the boundaries and parameters within which we live. There is something of a posthuman feel to Foucault's arguments here, with the technologies of power dictating social transformation (see Kittler 1999; Gane 2005; Hayles 1999). However we choose to frame these complex transformations, it is the visibility that these technologies of power afford that is central to the way that power operates.

Foucault uses the concept of the 'milieu' to develop these points a little further. This particular concept is useful because it brings together the circulation of population-based calculations with the power of technologies to render things and people visible. The concept of milieu then, despite only being discussed in very cursory terms by Foucault, is potentially useful for thinking through the relationship between metrics, visibility, and possibility. For instance, Foucault (2007: 21) outlines the concept by suggesting that 'the apparatuses of security work, fabricate, organize, and plan a milieu even before the notion was formed and isolated'. Crucially he then adds that 'the milieu, then, will be that in which circulation is carried out' (Foucault 2007: 21). The milieu is planned in the formation of the systems of calculation that were imposed. Governance here is not something that happens after they are measured, it is built into the design and structure of the very systems that produce those measurements. The things that are made visible and which are open to being limited or transformed are modelled into these very systems (see Chap. 3 for a discussion of this in relation to the values modelled in algorithms, which was also touched upon earlier in this chapter). The milieu is a planning-out of population based upon the way it will ultimately be measured. Foucault was unequivocal in seeing calculation as the means by which plans are realised. This, in Foucault's (2007: 21) accounts, is the product of 'the circulation of causes and effects that is targeted through the milieu'. The concept of the milieu here is used to promote the idea that lives and populations can be governed through calculation—with those calculations including planned causes, effects, and engineered outcomes. This may all be sounding a little like a version of Big Brother or even a type of technological determinism, but Foucault seems to be pointing at something rather more subtle. We need only to reflect on how work-based performance measures are designed to provoke action—to have causes and effects—to see that Foucault may have a point. Plus, his position on the milieu is not reductive or determinist; rather, it points to a historical moment in which population metrics were of a simpler and more rudimentary form. The milieu of today would be different, but it does suggest that the way that the milieu is made visible could still reveal some profoundly powerful plans or intended causes and effects. It is just that the calculations of today are going to be much more complex in the way that they bring the milieu into existence. The important point here though is that metrics can be used to realise plans. Which is to say that measures are often put in place with certain outcomes in mind. This concept of milieu is aimed at seeing how calculation is designed to achieve outcomes; the outcomes are designed into the measures. To encourage certain behaviours, practices, or actions, a suitable metric is needed to guide and provoke people towards a desired outcome. This is to try to anticipate the responses to metrics and to adapt them accordingly, so that they are gamed in desirable ways. We are left here though with some complex questions about agency—who has control of the data and how do people respond—that are going to be difficult to fully untangle (for a discussion of data and agency, see Kennedy and Moss 2015). This position perhaps places greater power in the hands of those planning the unfolding of metricisation, and suggests that the milieu will respond to those metrics in anticipated ways, but it is an important acknowledgement of the way that metrics are embedded into power structures. The attempt to calculate can be done with intended outcomes in terms of modelling and shaping behaviours. The visibility afforded by metrics can be targeted, with purpose.

The milieu that Foucault discusses inevitably relates closely to the question of biometrics that we discussed in Chap. 2. Frequently, the measures to which we are exposed are biopolitical, particularly as biometrics render visible the body—from DNA to fitness regimes. As the body is rendered visible, it lends itself to the production of corporeal truths about us. Pugliese (2010: 3) argues that 'biometric systems are inscribed as evidentiary technologies productive of "truth"'. In this case, the production of biometrics and the visibility of the body become enshrined as a kind of truth. Linking this back to Foucault, as we have just seen there is a sense that calculation produces or carves out reality (see also Adkins 2009: 336; Espeland and Stevens 2008: 417). In this case, these realities are based upon notions of evidence-founded truth. Pugliese continues by again reiterating the idea that value-based models of the social world are to be found in the design of the components of these systems. As he claims, 'embedded within the very infrastructural operations of the technology are a series of normative presuppositions and inscriptive categories (of race, (dis)ability, gender, age, class) that, because of their normative infrastructural status, cannot be named or rendered visible' (Pugliese 2010: 7). Thus these types of embedded values have visible outcomes whilst often remaining invisibly implicit within calculative systems (see Schinkel 2013: 1143). This creates an interplay between the empowering and disempowering forces of visibility. In some instances, it is powerful to be visible and disempowering to be invisible and in others it is powerful to be invisible and disempowering to be visible. This is a conundrum that needs to be explored through the type of work that Clare Burchill (2011) has done on transparency in political processes for instance. Burchill's work reveals the politics of this apparent transparency, with masses of public data being surrounded by accounts of the value of data transparency. The point is that these masses of public data, in spaces such as data.gov and data.gov.uk do not necessarily enable insights to be readily gleaned. The result is that the rhetoric and initiatives related to political transparency does not necessarily lead to an informed and empowered public. The interplay of visibility and invisibility is crucial in understanding metric power.

If we return to the key point, it has been argued that biometrics make the body visible and in so doing they overlay the body with 'calculatory grids' (Pugliese 2010: 8) against which it can be measured. As Pugliese (2010: 45) contends, biopower is 'effectively colonising the body, overlaying it with calculatory grids and geometrically inscribing it with formulae that will transform it into an object of knowledge and power'. These grids are made up of those constraining categories that organise the measures that are taken and which are argued to perpetuate existing norms, divisions, and prejudices. According to Pugliese (2010: 9), 'identificatory procedures of biometric technologies are fundamentally inscribed by tacit moralising assumptions, normative criteria and typological presuppositions'. In this formulation, metrics have built into them presuppositions and expectations that perpetuate existing ideas about the social world. These presuppositions are often invisibly laced into these systems, producing very visible judgements on the bodies to which those measures are being applied (the debates on 'somatechnics' and the body provide some examples of this, see Osuri 2009; Pugliese and Stryker 2009). Such an observation has led Pugliese (2010: 24) to argue that 'biometric systems are instrumental in the reproduction and maintenance of a political anatomy of the body premised on a mechanics of power that splits and fragments the hold between body, subject and identity'. These types of biometric technologies (see Chap. 2) are based upon a particular set of relations between visibility and knowledge, in which causal relations are determined in very specific ways and in which those calculations become facts about those bodies. As such, these new types of biometric technologies that quantify bodily properties and which enable unique identifiers are, as we have seen, to be 'situated within regimes of truth predicated on positivist ontologies of the visible' (Pugliese 2010: 157). These positive ontologies of the visible have a history that, as we saw in Chap. 2, takes us back through the history of social statistics and population metrics. To count is to know. This is a sentiment that is not limited to recent digital biometrics but is about the way that truths about people and populations are made from the visible properties of metrics. As the nature of biometrics have changed, so too has the way that the body is understood and approached. The body is caught up in these technological changes. As Pugliese (2010: 165) puts it, 'at the point of biometric enrolment, the body in question is at once a somatic singularity and a somatechnically mediated geocorpographically positioned figure enmeshed within normative and disciplinary networks of biopolitical power'. To understand how we are made visible requires us to understand the way our bodies enmesh with the wider values embedded in these systems of measurement. Norms are embodied through these techniques. As such, in more general terms:

> biopolitics requires a systematic knowledge of 'life' and of 'living being'… They make the reality of life conceivable and calculable in such a way that it can be shaped and transformed. Thus, it is necessary to comprehend the regime of truth (and its selectivity) that constitutes the background of biopolitical practices. One must ask what knowledge of the body and life

processes is assumed to be socially relevant and, by contrast, what alternative interpretations are devalued or marginalized. (Lemke 2011: 119)

Lemke's account is similarly positioned on the role of metrics in making bodies visible so that they might be held up to inspection for signs of normality. Here, measures enable the calculation of life and living, rendering them quantifiable. Individuals are then exposed, as Pugliese also argued, to regimes of truth via these biometric encounters. Lemke crucially poses the question about what knowledge of the body is deemed to be important, which is also to say which aspects of the body remain invisible, hidden, or concealed.

In such bodily focussed regimes of truth, it would seem that the key question is about what to make visible and what to leave invisible, these choices shape knowledge and shape outcomes. Effectively these questions of visibility around the body are choices around what is valued and what is not, or what is to be focussed upon and what is to be marginalised. It is in these choices about which types of metrics to promote and to circulate that the possibilities are carved out. This is clearly not a question that is limited to biometrics, but which might be opened up to encompass metrics more broadly conceived. How and what do metrics render visible? And what are the consequences of this visibility? This question might be asked of the broader 'regime of digital governance' (Day 2014: 133) to see what answers might be created. But visibility needs to be central to any such exploration. That is to say that when thinking about how metrics create, produce, or enact realities, we need to be focussing upon visibility as the means by which a measure becomes a tangible or empirical reality from which notions of difference, normality, and value might be gleaned.

There is then something of a 'symbiotic relation' between transparency and secrecy (Ruppert 2015: 146), with disclosures potentially being 'buried under volumes of data thus rendering it opaque'. This conjures the image of the needle in a haystack. This is a significant part of the relations between the visible and invisible. With these issues of visibility in mind, it is important to consider the various ways that bodies, actions, and collectives are rendered visible by metrics. The politics around visibility are not straightforward though. Despite the long history outlined in Chap. 2, there is now an even more complex interplay between the visible and invisible. This complex interplay of the visible and the invisible then defines the interplay between empowerment and disempowerment. The point is a slippery one, but it is to say that in understanding the possibilities that circulating metrics create, understanding the interplay of visibility and invisibility will be crucial. What is visible is what is possible. What is invisible is what is unknown, unknowable, unlikely, discouraged, or marginalised. Similarly, though, being visible is to be exposed, to be vulnerable, and to be judged.

Probability and the Politics of Possibility

We have already seen some suggestion that measurement and calculation have a powerful ordering and limiting presence. Through Elden's (2006) work on Heidegger, for example, we saw that calculation was argued to shape perceptions and form limits. It is possible to reflect then on how measurement and metrics produce the social world in different ways as they come to intervene in the perception, performance, and practices of individuals and groups. Louise Amoore's (2013) work presents a particularly compelling case for a more refined understanding of what she calls 'the politics of possibility'. The crucial shift, for Amoore, is in the move from the use of statistics and metrics in the analysis of probabilities or likelihoods and towards their use in the analysis of possibilities. The argument here is that there are occasions and environments today in which metrics are used to locate a variety of potential outcomes or possibilities—which in turn then produce realities and outcomes. For Amoore (2013: 1), 'the idea that uncertain futures—however probabilistically unlikely—be mapped and acted upon as possibilities has captured the Zeitgeist'. The power of data rests, in this view, with its ability to tell various organisations what is possible. The result is that predictions and anticipatory actions can be derived from metrics. In the case of security, for instance,

> [t]he data that are mined, integrated, and analyzed within e-Borders are of a specific type—they are designed to look for not only the settled probabilities

of future risk (such as names on watch lists or no-fly lists) but also for the possible risks amid an array of future uncertainties. (Amoore 2013: 1)

Risk is assessed through the possibilities that are produced from the data. Possible risks are found in the data. Thus possibility is the focus, rather than likelihood. As Amoore (2013: 3) claims, 'it is precisely across the gaps of what can be known that new subjects and things are called into being'. These future uncertainties become part of the present and offer new forms of knowledge about the social world that fill these gaps. This though goes beyond the sphere of security in the narrow sense. Crucially Amoore argues:

> Thus it is that there is an intensifying resonance across spheres of economy and security, an infiltration of each one into the other, such that a moving complex emerges—a complex of the governing of emergent, uncertain, *possible* futures. The point of resonance on the horizon, I propose, is precisely a horizon of possible futures, arrayed in such a way as to govern, to decide, or to act in the present. (Amoore 2013: 5)

Economic and security decisions, according to Amoore, interweave as these new data assemblages facilitate the analysis of possible futures. The result is a vast array of imaginable outcomes, of which some are highlighted and acted upon. In other words, some of these projected futures become part of decision making and action. Contemporary decisions and judgements then can be predicated on this 'the horizon of possible futures' (Amoore 2013: 6).

Resonating with observations made by Will Davies (2014), Amoore sees this as being a space in which the analytical infrastructure of consultation expands. According to Amoore (2013: 6), what we find is 'the proliferation of *consulting*—a way of thinking, ordering, calculating, and acting in the world—rather than primarily the actions of consultants as identifiable agents'. Again, we see the importance of what Elden (2006) described as a 'mode of thinking' coming to dominate decision making. These decisions, Amoore adds, are often about risk. As she explains:

> The politics of possibility sees a seemingly limitless series of bodies, populations, spaces, buildings, financial transactions, tickets, movements, shapes, and forms divided and fractionated according to degrees of risk. (Amoore 2013: 7)

This is a politics based on what can be measured, what can be assessed for risk, and how this can then shape outcomes. As such, a version of the future becomes part of current decision making and action. The outcomes are shaped by the visions of the future that are drawn upon. This is significant because, according to Amoore (2013: 7), 'risk technologies have, at their heart, a particular relationship to the future. They hold out the promise of managing uncertainty and making an unknowable and indeterminate future knowable and calculable'. As a consequence this particular set of relations with the future, these particular visions of what is possible, become reified in the actions they inform. The unknown becomes something tangible that can be measured and acted against. Thus circulating measures define what is seen to be possible. This is an approach to risk that is based upon anticipation. Amoore contends that:

> [t]he specific modality of risk…acts not strictly to prevent the playing out of a particular course of events on the basis of past data tracked forward into probable futures but to preempt an unfolding and emergent event in relation to an array of possible projected futures. It seeks not to forestall the future via calculation but to incorporate the very unknowability and profound uncertainty of the future into imminent decision. (Amoore 2013: 9)

Anticipation, informed by metrics, becomes a potential reality in the present. The nature of that anticipation and of the visions from which it is drawn is powerful in shaping choices and outcomes. Authority comes to derive from the ability to anticipate possibilities. Visions of the future, or these 'lines of sight' of possible horizons, as Amoore refers to them, provide legitimacy whilst also having powerful constitutive effects. This goes beyond the human responses to measures captured by Espeland and Sauder's (2007) discussion of 'reactivity'; this is about automated systems responding to metrics and carving out possible futures. As a result, Amoore claims, 'scenario planning, risk profiling, algorithmic modelling, information integration, and data analysis become the authoritative knowledges of choice' (Amoore 2013: 9). Or, as Amoore later adds, 'the risk techniques that flourish on the horizon of possibilities…are not those of auditing, accounting, statistics, actuarial science, and probability but instead are those of consulting, screening, remote tracking, biometric identifying, and algorithmic profiling' (Amoore 2013: 13). Knowledge of possible futures becomes the valued knowledge of the time. This knowledge of possible scenarios takes a particular form, in that it is based on certain types of data and certain analytics, meaning that it presents certain possibilities above others. These imagined scenarios then become the means by which current possibilities are derived.

Clearly then the key distinction that Amoore makes, and which is crucial to the position developed in her book, is between probability and possibility. It is the shift that is at the centre of the arguments about the emergence of the politics of possibility. Amoore (2013: 12) argues that:

> the mode of risk that is flourishing across the horizons of contemporary economy and security operates according to a possibilistic logic. It does not deploy statistical probabilistic calculation in order to avert future risks but rather flourishes in conditions of declared constant emergency because decisions are taken on the basis of future possibilities, however improbable or unlikely. (Amoore 2013: 12)

The calculations of what is probable are usurped then by the urgency of what is possible. The concern here is not with what is probable or the factoring of chance, which we focussed upon in Chap. 2, but rather it is with scoping out the various possibilities and using these to anticipate. Thus, we see a shift here in the power of metrics away from the lineage of probability and towards something different—what we talked about in terms of the appreciation of chance changes here to something that is not directly concerned with likelihood. As Amoore further emphasises, 'risk in the mode of possibility rather than strict probability, does not govern by the deductive proving or disproving of scientific and statistical data but by the inductive incorporation of suspicion, imagination, and pre-emption' (Amoore 2013: 10). The shift is from the analysis of likelihood to an analytics of the possible and to the pursuit of pre-emptive action. Decisions then are taken not even on a probable imagined future but on future possibilities as imagined through the metrics. Notions of risk are central to this. This is to use metrics to make decisions that subvert a future event. This, for Amoore, is an important distinction to make. The politics of possibility is based upon this shift, and is also indicative of wider political conditions. The sense of emergency becomes the means by which this shift is made politically and culturally viable it would seem. This is decision making based on 'intuitive' or 'speculative' forms of knowledge, which are modelled by 'data-led algorithms' (Amoore 2013: 10). The particular mode of reasoning associated with calculation and probability that we encountered in Chap. 2 is now reformulated into a new mode of reasoning based on pre-emption. This is a milieu that doesn't need to have its future planned, as we've discussed in this chapter; rather, it is a milieu in which future planning circulates back into the present.

So how does this speculative knowledge of future possibilities shape what is possible? As the above has already suggested, one key influence emerges as future possibilities become part of current decision making, thus shaping social outcomes, activities, planning, and responses. As Amoore (2013: 75) puts it, 'in its derivative form, the risk calculus folds future possibilities into present decisions, making present the future consequences of events that may or may not come to pass'. Data become all-important as analytics are used to seek the possibilities that they present. It is not clear what data will be used or which possibilities might be important. This leads, Amoore (2013: 89) notes, to 'a horizon of possibility where all incoming data may become significant in the future and, therefore, must be arrayed in view'. This is not about individual projections but horizons of ranging possible scenarios based upon the pursuit of vast accumulations of data about people. It is a mode of thinking that might represent a mutation from those discussed in Chap. 2—it is not just a calculative mode of thinking but also an anticipatory, predictive, or analytic mode of intervention. As a result, the use of data to form such horizons means that 'as objects, the data appear technoscientific, hollowed out or shell-like…as a thing, it also lives, moves, has unintended consequences' (Amoore 2013: 154; which relates to our earlier discussion of metrics and the erosion of narrative in Chap. 3). For Amoore, data are seen as objects that are used to create meaning, but these meanings create consequences and outcomes that cannot be foreseen. Even errors, mistakes, and problematic predictions can become realities (as has also been pointed out by Hayles 1999)—with stories of computer errors that become very real problems for people, with instances such as speeding fines and foreclosures a fairly regular concern in the popular media (Naughton 2015). As Amoore (2013: 156) explains, 'the politics of possibility appears to incorporate every element, even its own mistakes and errors are folded back into the capacity to write new code and locate new correlations'. These circulatory processes result in all aspects of the data becoming part of the modelling processes and therefore shaping the possibilities that are generated and the decisions that are made. So, this is about current decisions, but it is also about the way that those data and decisions are modelled into the systems that are used in future decision making. These systems are self-referential in the futures they imagine.

Finally though, in emphasising the role of circulating measures in affording certain possible outcomes, Amoore, like Elden (2006), points to the limiting powers of these forms of analytics. Amoore (2013: 157) suggests that 'where the politics of possibility actualizes the unknown in material forms—risk scores, visualized networks, decisions trees, protocols—the potential of people and objects is never fully actualized and, therefore, never meaningfully incorporable'. The important point here is that the projected horizons of possibility that Amoore discusses, as produced by data and algorithmic systems, comes to define what is actualised and what becomes a reality. For Elden the limiting factors are in calculation, and for Amoore the limiting factors are to be found in the projection of possibilities. Metric power then works by shaping what is known through calculation and what predictions are made. One of the circulations typical of metric power is between the future and the present. The metrics are used then to shape what is possible through the imagined outcomes that they afford.

Conclusion

When thinking about metric power it is important to not just think of circulating metrics, we also need to think about the way that these circulating metrics shape either what is possible or what might be conceived to be possible. In Chap. 2, we talked briefly about the role of probability in the history of the measurement of the social world. We have ended this chapter with some reflection on the move from a focus on probability to an interest in possibility. This instantly shows how it is not just the measurements that matter, it is what they are used for and the type of understandings they are used to promote and legitimate that counts. When thinking of the relations between metrics and possibility then, Amoore's notion of the 'politics of possibility' should be a central touchstone for future work. It is crucial that such a shift should be interrogated and the very notion of metric-based possibilities should be the nub of intense, rigorous, and relentless examination. Amoore's work deals directly with the very notion of possibility itself and with how the possible outcomes and future are imagined. This indicates that circulating metrics feed into the conditions of possibility in other ways—a number of which link to Amoore's work but also bring in broader debates in the social sciences.

More broadly, this chapter has demonstrated how metrics enact and craft possibilities in a range of registers. This chapter has reflected upon the way that metrics define possibility in terms of the fostering of inequality, the definition of value and worth, and in the carving of visibility. Each of these shows that in terms of understanding metric power we need to bring into focus what Bauman (2011) calls 'collateral damage'. That is to say that we need to think about the way that metrics circulate through the social world defining and affording what is possible for individuals and their lives. This chapter would suggest that this is a pressing question for social scientists of all types. Many sectors of the social world are now being reworked and reanimated by metrics, so whatever aspects of the social world we might be interested in exploring metric power is likely to be operating upon and within it. This chapter has only begun to touch upon the complex ways in which metrics feed back into the conditions of possibility—drawing on questions of inequality, value, visibility, and imagined futures—there are many more to be explored.

What this chapter suggests is related to Ronald Day's (2014: 135; italics in the original) conclusion that the '*neoliberal documentary society is governed by means of collectively managed self-adaptation, afforded by documentary mediation*'. In this chapter, we have seen how metrics play a role in adaptation, with power central to how this collective management works to maintain, strengthen, or justify new types of inequality, to define value or worth, and to make the selections central to affording visibility or invisibility. Metrics, therefore, are playing an important and functional part in how such power formations function. Crucially, metrics, as documentary and indexed components of the social world, mediate possibility by defining and shaping what is possible and what is seen to be possible. It is only when we place these questions of the metricisation of possibility alongside the earlier insights into measure and circulation that we might begin to see the inescapable strength of metric power.

References

Adkins, L. (2009). Feminism after measure. *Feminist Theory, 10*(3), 323–339.

Adkins, L., & Lury, C. (2012). Introduction: Special measures. In L. Adkins & C. Lury (Eds.), *Measure and value* (pp. 5–23). Oxford: Wiley-Blackwell.

Amoore, L. (2013). *The politics of possibility: Risk and security beyond possibility.* Durham, NC: Duke University Press.

Atkinson, R., & Burrows, R. (2014). A city in thrall to capital? London, money-power and elites. *Discover Society, 15.* Accessed July 6, 2015, from http://discoversociety.org/2014/12/01/a-city-in-thrall-to-capital-london-money-power-and-elites/

Badiou, A. (2008). *Number and numbers.* Cambridge: Polity Press.

Bauman, Z. (2007). *Consuming life.* Cambridge: Polity Press.

Bauman, Z. (2011). *Collateral damage: Social inequalities in a global age.* Cambridge: Polity Press.

Beer, D. (2013). *Popular culture and new media: The politics of circulation.* Basingstoke: Palgrave Macmillan.

Beer, D. (2015b). Productive measures: Culture and measurement in the context of everyday neoliberalism. *Big Data and Society, 2*(1), 1–12.

Bennett, T., Savage, M., Silva, E. B., Warde, A., Gayo-Cal, M., & Wright, D. (2009). *Culture, class, distinction.* Abingdon: Routledge.

Brown, W. (2015b). *Undoing the Demos: Neoliberalism's stealth revolution.* New York: Zone Books.

Burchill, C. (2011). Introduction to 'secrecy and transparency': The politics of opacity and openness. *Theory Culture and Society, 28*(7–8), 7–25.

Crossley, N. (2009). The man whose web expanded: Network dynamics in Manchester's post/punk music scene 1976–1980. *Poetics, 37*(1), 24–49.

Davies, W. (2014). *The limits of neoliberalism.* London: Sage.

Davies, W. (2015a). *The happiness industry: How the government and big business sold us well-being.* London: Verso.

Day, R. E. (2014). *Indexing it all: The subject in the age of documentation, information, and data.* Cambridge, MA: MIT Press.

Dodd, N. (2014). Piketty symposium. *British Journal of Sociology, 65*(4), 589–747.

Doria, L. (2013). *Calculating the human: Universal calculability in the age of quality assurance.* Basingstoke: Palgrave Macmillan.

Elden, S. (2006). *Speaking against number: Heidegger, language and the politics of calculation.* Edinburgh: Edinburgh University Press.

Espeland, W. N. (1997). Authority by the numbers: Porter on quantification, discretion, and the legitimation of expertise. *Law and Social Inquiry, 22*(4), 1107–1133.

Espeland, W. N., & Sauder, M. (2007). Rankings and reactivity: How public measures recreate social worlds. *American Journal of Sociology, 113*(1), 1–40.

Espeland, W. N., & Stevens, M. L. (2008). A sociology of quantification. *European Journal of Sociology, 49*(3), 401–436.

Foucault, M. (2007). *Security, territory, population: Lectures at the Collège de France 1977–1978.* Basingstoke: Palgrave Macmillan.

Foucault, M. (2008). *The birth of biopolitics: Lectures at the Collège de France 1978–1979.* Basingstoke: Palgrave Macmillan.

Gane, N. (2005). Radical post-humanism: Friedrich Kittler and the primacy of technology. *Theory Culture and Society, 22*(3), 25–41.

Gordon, C. (1991). Governmental rationality: An introduction. In G. Burchill, C. Gordon, & P. Miller (Eds.), *The Foucault effect* (pp. 1–51). Chicago: The University of Chicago Press.

Hacking, I. (1990). *The taming of chance.* Cambridge: Cambridge University Press.

Hacking, I. (1991). How should we do the history of statistics? In G. Burchill, C. Gordon, & P. Miller (Eds.), *The Foucault effect* (pp. 181–195). Chicago: The University of Chicago Press.

Halpern, O. (2014). *Beautiful data: A history of vision and reason since 1945.* Durham, NC: Duke University Press.

Hayles, N. K. (1999). *How we became posthuman: Virtual bodies in cybernetics, literature, and informatics.* Chicago: The University of Chicago Press.

Hearn, A. (2008). 'Meat, mask, burden': Probing the contours of the branded 'self'. *Journal of Consumer Culture, 8*(2), 197–217.

Hearn, A. (2010). Structuring feeling: Web 2.0, online ranking and rating, and the digital 'reputation' economy. *Ephemera, 10*(3–4), 421–438.

Higgins, V., & Larner, W. (2010a). Standards and standardization as a social scientific problem. In V. Higgins & W. Larner (Eds.), *Calculating the social: Standards and the reconfiguration of governing* (pp. 1–18). Basingstoke: Palgrave Macmillan.

Higgins, V., & Larner, W. (2010b). From standardization to standardizing work. In V. Higgins & W. Larner (Eds.), *Calculating the social: Standards and the reconfiguration of governing* (pp. 205–218). Basingstoke: Palgrave Macmillan.

Kennedy, H. & Moss, G. (2015). Known or knowing publics? Social media data mining and the question of public agency. *Big Data and Society*, 1–11. doi: 10.1177/2053951715611145.

Kitchin, R. (2014a). *The data revolution: Big data, open data, data infrastructures & their consequences*. London: Sage.

Kittler, F. (1999). *Gramophone, film, typewriter*. Stanford, CA: Stanford University Press.

Konings, M. (2015). *The emotional logic of capitalism: What progressives have missed*. Stanford, CA: Stanford University Press.

Lemke, T. (2011). *Bio-politics: An advanced introduction*. New York: New York University Press.

MacKenzie, D. (2015, May 21). On 'Spoofing'. London Review of Books, p. 38.

Mair, M., Greiffenhagen, C., & Sharrock, W. (2015). Statistical practice: Putting society on display. *Theory, Culture and Society*. Online first. doi: 10.1177/0263276414559058.

Mckenzie, L. (2015). *Getting by: Estates, Class and Culture in Austerity Britain*. Bristol: Policy Press.

Miller, P., & Rose, N. (2008). *Governing the present*. Cambridge: Polity Press.

Moor, L., & Lury, C. (2011). Making and measuring value: Comparison, singularity and agency in brand valuation practice. *Journal of Cultural Economy*, 4(4), 439–454.

Naughton, J. (2015, August 16). Why its full speed ahead to being ruled by computers. *The Guardian*. Accessed August 19, 2015, from http://www.theguardian.com/commentisfree/2015/aug/16/full-speed-ahead-being-ruled-by-computers

Oksala, J. (2013). Feminism and neoliberal governmentality. *Foucault Studies, 16*(1), 32–53.

Osuri, G. (2009). Necropolitical complicities: (Re)constructing a normative somatechnics of Iraq. *Social Semiotics, 19*(1), 31–45.

Pasquale, F. (2015). *The black box society: The secret algorithms that control money and information*. Cambridge, MA: Harvard University Press.

Peck, J. (2012). Austerity urbanism: American cities under extreme economy. *CITY, 16*(6), 626–655.

Piketty, T. (2014). *Capital in the twent-first century*. New York: Harvard University Press.

Porter, T. M. (1995). *Trust in numbers: The pursuit of objectivity in science and public life*. Princeton, NJ: Princeton University Press.

Pugliese, J. (2010). *Biometrics: Bodies, technologies, biopolitics*. London: Routledge.

Pugliese, J., & Stryker, S. (2009). The somatechnics of race and whiteness. *Social Semiotics, 19*(1), 1–8.

Roy, A. (2014). *Capitalism: A ghost story*. London: Verso.

Ruppert, E. (2015). Doing the transparent state: Open government data as performance indicators. In R. Rottenburg, S. E. Merry, S. J. Park, & J. Mugler (Eds.), *The world of indicators: The making of governmental knowledge through quantification* (pp. 127–150). Cambridge: Cambridge University Press.

Savage, M. (2013). The 'Social life of methods': A critical introduction. *Theory Culture and Society, 30*(4), 3–21.

Savage, M. (2015). *Social Class in the 21st Century*. London: Pelican Books.

Scharff, C. (2014). Gender and neoliberalism: Exploring the exclusions and contours of neoliberal subjectivities. *Theory, Culture and Society*. Accessed August 20, 2015, from http://theoryculturesociety.org/christina-scharff-on-gender-and-neoliberalism/

Schinkel, W. (2013). The imagination of 'society' in measurements of immigrant integration. *Ethnic and Racial Studies, 36*(7), 1142–1161.

Skeggs, B. (2005). The making of class and gender through visualizing moral subject formation. *Sociology, 39*(5), 965–982.

Skeggs, B. (2014). Values beyond value? Is anything beyond the logic of capital? *The British Journal of Sociology, 65*(1), 1–20.

Skeggs, B., & Yuill S. (2015). Capital experimentation with person/a formation: How Facebook's monetization refigures the relationship between property, personhood and protest. *Information, Communicaton and Society*. Online first. doi: 10.1080/1369118X.2015.1111403.

Sociological Review. (2015). Sociologies of class: Elites (GBCS) and critiques. *Sociological Review, 63*(2), 205–549.

Tyler, I. (2013). *Revolting subjects: Social abjection resistance in neoliberal Britain*. London: Zed Books.

Tyler, I. (2015). Classificatory struggles: Class, culture and inequality in neoliberal times. *The Sociological Review, 63*(2), 493–511.

Tyler, I., & Bennett, B. (2010). 'Celebrity chav': Fame, femininity and social class. *European Journal of Cultural Studies, 13*(3), 375–393.

Venn, C. (2009). Neoliberal political economy, biopolitics and colonialism: A transcolonial genealogy of inequality. *Theory Culture and Society, 26*(6), 206–233.

White, M. D. (2014). *The illusion of well-being: Economic policymaking based on respect and responsiveness.* New York: Palgrave Macmillan.

5

Conclusion: The Intersections and Imbrications of Metric Power

Espeland and Stevens (2008: 411) have acknowledged that in 'a world saturated with numbers, it is easy to take the work of quantification for granted'. We certainly should resist such a temptation. The world is indeed saturated by metrics, but it would be a mistake to overlook their power simply because of their familiarity. Espeland and Stevens' argument is that we need sociological insights that think about the background work, knowledge, and expertise that are required to make these metrics so powerful. We need, in short, to think of these metrics in sociological terms. This book argues that we should in fact see metrics as being central to the ordering, division, and construction of the social world today. We are governed, managed, and corralled by metrics; they act upon us and through us. We should see this in its long historical context (see Chap. 2), but the role and presence of metrics have undoubtedly escalated in recent years. We might even conclude that we have moved from the threat of intermittent measures acting upon us—in the form of 'special measures'—to forms of power that act through our constant exposure to those metrics. Special measures have become ordinary (see Beer 2015c).

Metric power is a concept that is intended to focus attention on the relations between measurement, circulation, and possibility in order to

© The Editor(s) (if applicable) and The Author(s) 2016
D. Beer, *Metric Power*, DOI 10.1057/978-1-137-55649-3_5

extend our understanding of the linkages between metrics and power. As Wendy Espeland (1997: 1120) has argued, 'power, self-interest, and informal knowledge will always mediate the use of number'. We might have to revisit Espeland's statement in light of automated circulatory systems and the potential usurping of informal knowledge by complex anticipatory knowing, but nevertheless her point certainly pertains today. It is only by understanding how measurements circulate into the world, shaping what is possible and what is seen to be possible, that we might understand the power dynamics of metrics. It is also in these relations that we might unpick the power dynamics behind the so-called big data. It is important to see metric power in context. That is to say that it is important to see it as part of broader historical and political forces.

If we take a step back though, and perhaps resist some of the more epochal (Savage 2009) tendencies that are tempting when using a term like metric power, Foucault (2002b: 284) once said that for him 'power is what needs to be explained, rather than being something that offers an explanation'. The problem was that the question of power, he pointed out, is something that is difficult to deal with. It is the thing that requires analytical attention and with which we need to 'grapple'. In this book I've tried to work towards some explanation of power as it operates through metrics and to grapple with the questions this presents. Writing around 25 years ago, Nicholas Rose (1991: 673) noted that 'numbers have an unmistakable power in modern political culture' and that even the 'most casual reader of newspapers or viewer of television is embraced within the rituals of expectation, speculation, and prognostication that surround the public pronouncement of politically salient numbers'. It is the unmistakable power to which Rose refers here that we still need to continue to grapple with, particularly as the newspapers and television, to which he refers, have been appended with mobile devices, social media, GPS, Wi-Fi, and the many opportunities that these types of media bring to capture and disseminate metrics and metric-based thinking—in our labour, our consumption, our social relations, and our movements. This power associated with numbers has changed, and so we must grapple with it again. Beyond this, given the current context, we must redouble our efforts. The analytical gaze associated with metrics is something to which we will need to pay extra attention. Porter has warned that it is easy to look at where metric-based indicators intimate, to stare in the direction in which those indicators gesture, leading us to look away from the actual sites of contested power. As Porter (2015: 36) puts it, we can end up looking 'so intently to where it points that we neglect the reality and power of the indicator'. We need then to resist only looking to where we are being pointed, and to give some attention to the very thing that is gesturing to us.

The concept of metric power, as briefly described in the introduction of this book, is intended to sensitise us to the role of metrics in the performance of the social and the intricacies of the numerical governance of individuals and populations. More than this though, it is a concept that is intended to focus attention upon the relations between *measurement*, *circulation*, and *possibility*. These are the intersections and imbrications that are central to the functioning and reach of metric power. It is in an analysis of these relations that we are most likely to develop a clearer idea of the linkages between metrics and power or to understand the part that metrics play in power relations, dynamics, and structures. It is also here that we might grasp the continuing intensification of metrics in the social world.

If, as we are led to believe, and with some compelling force, competition is central to the organisation of the social world under the forces of neoliberalisation, then understanding metrics becomes even more pressing. Statistics and measurement have a long history, as we have briefly seen in Chap. 2, but it is interesting to consider what happens when this history converges with processes of neoliberalisation and as measurement takes on a particular form in the enhancement and expansion of competition. Metrics provide the mechanisms by which competition is able to be performed. The advancement of metrics has certainly unblocked the pathways of neoliberal governance today, and the taps are now gushing. Measures are needed for the differentiation required by competition. The argument would go that with the rise of digital media infrastructures of various types, a whole new apparatus of measurement has emerged from a long genealogical line of development. The scope of measurement has increased and so metricisation has intensified. With the rise of mobile devices, smartphones, apps, and many other tracking technologies and 'logjects' (Dodge and Kitchin 2009), so too the range of ways that we can be measured, how often and to what extent has increased. This is an increase in the desire to measure complemented by an increase in the ability to measure. We may have seen this materialise in the visualisation of our performance at work, or it may be the way that we use a smartphone app to measure our bike speed compared to other users, or how we follow the data about the number of retweets or reblogs we have had, the quantification of our bodies in healthcare, our credit scores or insurance risk profiles (be they individual or national, see Farlow 2015: 228–234), how we are ranked in league tables, or perhaps just the number of likes or friends we have on Facebook. The list goes on. We could add more and more detail, but this is suggestive of both our own interest in being measured and competing, combined with the interests of others in making us compete. These desires and agendas are then aligned with an industry of analytics and a deepening of the metric assemblage. In the previous three chapters we dealt with measurement, circulation, and possibility as separate entities, albeit in some sort of iterative order. What I would like to do in this brief conclusion is to think about how the findings in these previous chapters might come together to inform a notion of metric power and the relations that underpin it.

We might start with a consideration of the type of power we are dealing with. Miller and Rose (2008) have offered some insights into a decentralised form of power that might be of use at this juncture. They focus upon developing an understanding of 'power without a centre', moving instead to power with 'multiple centres'. This is power that is 'productive of meanings, of interventions, of entities, of processes, of objects, of written traces and of lives' (Miller and Rose 2008: 9). There is certainly something that we might take from such a position. This provides a helpful resource for reflecting on the form that metric power might be taking, as it operates around various 'centres' or as it produces meanings, affords interventions, and traces lives. But I'd like to try to develop something very specific within this conclusion, it is a position that might highlight a particular way in which power works through metrics—a particular type of power that appears to mutate out of some of these broader observations and which continues to gain traction, confidence, and appeal. As I have suggested, one way of thinking beyond these difficulties in grasping metric power is to focus our attention upon the relations between the three key areas that I have covered in this book. From the previous discussions of measurement, circulation, and possibility, we can perhaps identity a series of cross-cutting themes that may then come to inform and flesh out the power of metrics. I have covered other important issues in this book, but I have selected the most significant of these in order to sketch out the potential of the concept of metric power. These cross-cutting themes help in understanding how measurement, circulation, and possibility relate to one another and show the types of issues that might be revealed should we pay closer attention to these crucial relations.

First, we have seen that metric power is based upon the creation and maintenance of limits and parameters. Metric power operates through the carving of liminal boundaries and by constraining the pursuit of certain possibilities. This is a form of control through limit, a form of power that operates by shaping edges to what can be known and by channelling activity in certain directions through its presentation of pre-set constraints that shut down options, choices, and movements. It places a kind of measure upon the imagination and locks down potential. Metrics create rigid numerical edge-points, or final boundaries that set the score and become the rules and norms of the game.

Second, and allied with the above, metric power is based upon what it renders visible and invisible. Metrics allow some things to be seen and others to be hidden. As a result, certain things become important and others become marginal. This is not a simple set of relations, visibility and invisibility can be used to distribute power in different and variegated ways. Metrics have the power to make visible or to leave invisible. Metrics can be used to expose or conceal, to highlight or obfuscate, to illuminate or shade. In some instances visibility is to empower; in others, it is disempower. Similarly, to make invisible can be disempowering, but it might also be used to keep powerful secrets or to manipulate possibilities. As such, it is in the politics of visibility that metrics come to define what is known and what is knowable. This power through visibility also translates directly into value. Certain practices, objects, behaviours, and the like become visible and thus gain value, whilst others are not measured, noticed, or ultimately valued. Metric power works through this complex interplay between the visible and the invisible.

Third, and perhaps most obviously, metric power works by ordering, sorting, and categorising. We already know the power of categorisation (see e.g. Foucault 2002a; Bowker and Starr 1999; and for a recent discussion see Tyler 2015), metrics contribute to the heightening of ordering processes and their instantiation in social life. Metric power works by increasing the possibilities for categorisation and by increasing the possibilities for their translation into the social world. Metrics have the capacity to order and to divide, to group or to individualise, to make-us-up and to sort-us-out (on sorting out see Burrows and Ellison 2004). These orders, divisions, and categories pre-set opportunity and place those lives within patterns of judgement.

Fourth, metric power works by prefiguring judgements and setting desired aims and outcomes. As such, measurements do not only capture they also produce. It is important to remember that we are looking here at a way of thinking as well as a way of measuring. The use of metrics is often based upon desired outcomes, with the measures being used to try to move towards those ends. As such, metrics are based upon models of the world, these models, such as those used in algorithm design, have the potential to become realities in their own right and to fulfil their own prophecies, to perpetuate disparity and so on. One way this works is that metric power can be used to bring the future into the present. As we have seen described in Louise Amoore's (2013) key work (see Chap. 4), metrics can be used in setting out horizons, savannas, vistas, and imagined futures and then using them in current decision-making processes. But what are these models of the world? We might briefly turn again to Foucault's (2008: 148) contention that the:

> multiplication of the 'enterprise' form within the social body is what is at stake in neo-liberal policy. It is a matter of making the market, competition, and so the enterprise, into what could be called the formative power of society.

The model here is one of enterprise (see also Dardot and Laval 2013: 259; and for a discussion of the relation between freedom and free enterprise see Harvey 2005: 37). McNay (2009) also points to the 'self as enterprise' as a model for questions around autonomy, resistance,

weight, and scope. As the previous chapters have made clear, the rise of calculation and measurement has been an ongoing process, but what we are seeing here is an escalation in the opportunities to measure and in the scope and reach of metrics of different types. As such, it is important to think of the way that metrics and power come to interweave with one another. Indeed, this is increasingly pressing as metrics come to hold such sway in how the social world is organised across virtually all social spheres with international as well as personal consequences—from grand calculations of national GDP or state deficits through to very personal measures of our emotions or bodily activities. Metrics may, in some instances, be misrepresentative, misleading, or even inaccurate, yet they have powerful sway. The way that metric-based indicators simplify the complexity of the social world may be one reason for us to question their representativeness (Espeland 2015: 61), but this doesn't make them any less forceful. A key component or property of metrics in affording this sway is that they are seen to be neutral and objective. Metrics become very much a material reality once they are drawn upon or adopted into social and cultural practice. The example of spoofing in 'anonymous and electronic' financial markets shows how metrics can be used to create realities or to manipulate those realities. Donald MacKenzie (2015) has described how fake bids can deceive markets and give the impression of high levels of demand, which outstrip supply and the price changes, the point being that realities can be manipulated and remade in response to metrics and their deployment, especially where algorithms are making the decisions. Thus, metric power is a complex and decentred form of power that can work in surprising ways by engineering realities that can then be re-crafted, manipulated, and redrawn.

The pursuit of a more refined understanding of metric power cannot end here. It is actually tied up with some of the most pressing questions of our time. Metric power is at the forefront of our defining political transformations, but it is also—and this makes it even more complex—at the forefront of how that metric-defined world is researched and understood by social researchers and commercial analytics providers alike (for more on commercial analytics see Burrows and Gane 2006). We are left then to ask how measurement might reconfigure the social world itself and also how, as a part of this, metrics are the resource through which we know that world.

One obvious question we might ask concerns our potential resistance to metric power. We have seen that numbers are appealing and hard to resist. We have also seen the rise of metrics is not a sudden change but one which can be seen to have routes to the seventeenth century and the increase in the use of numerical measures through the nineteenth and twentieth centuries (see Chap. 2). In other words, it is something with long and tangled roots in the social world in which we live. Desrosières provides us with a clue as to where we might start with such resistance. He claims that 'inscribing a measurement in a system of negotiations and stabilized institutions (for example, through rules of indexing) can provide arguments for denying the objectivity and consistency of certain statistical indicators' (Desrosières 1998: 332). Desrosières suggests that this approach can show that such measures and approaches are not fixed. It seems unlikely, given what has already been concluded, that merely raising questions about objectivity and consistency will carry much weight or will have the momentum to topple any unwelcome exercising of metric power. As we saw through the work of Stuart Elden (2006) in Chap. 2, we would need to focus on the ways or modes of thinking behind calculation rather than trying to question the numbers themselves. The faith in metrics is likely to be too engrained or entrenched, with too much invested in it by those in positions of power and with the usually self-reinforcing properties that metrics bring. One possibility, similar to that suggested by Bev Skeggs (2014), might be to find the ruptures that occur when we locate things that are hard to measure. Take this account of immeasurability provided by Doria (2013: 1):

> Calculation tends to encompass 'objects'—such as emotions, affects, psychological well-being, social relations, intellectual labour—which historically have been at its periphery. What appears before us is an unlimited calculative proliferation ever more directly involving human life, down to its most 'incalculable' aspects; and precisely when we are faced with the calculation of the incalculable human, the awkward question of the unconditional character of calculation is projected as a disturbing shadow.

Metrics are often used to try to capture or manipulate things that are not comfortably measured or which we might argue can't be measured (for a discussion of this in the case of well-being see White 2014). These then become objects. This poses a kind of awkward question for Doria. The proliferation of attempts to calculate the human can perhaps be questioned by highlighting and making increasingly visible the problems of measure that occur when incalculable things about us are subject to measure (we will focus on affect, for instance, in Chap. 6). Perhaps it is in the attempts to metricise such things that we will find the purchase by which more pertinent questions might be asked about the advancement of the unlimited measurement of life to which Doria refers.

The central problem is that acts of resistance to metric power can easily get locked into the logic of the metrics themselves and the mode of thinking that informs them. When debates, observations, findings, and judgements are framed by metrics, it might seem that the only way to respond is with alternative interpretations of the numbers on the table. Metrics are simply too weighty and convincing to simply be dismissed or undermined; that is the problem for those wishing to resist their logic. Any other response, as Day (2014) has noted, is likely to be dismissed as naive, sour grapes, or illogical. As such, one inevitable possibility that we have to consider is that resistance might have to come from within the logic and rhetoric of metric power. To be empowered is to know and understand our own metrics—and to understand the thinking that is behind the way that they are being deployed. To do this, we might get to know the different ways that we are measured, plus we might make sure that we know how to interpret those metrics so that we can offer our own alternative accounts. This would be to find our own measures and to have them ready as a resource with which to challenge any negative or damaging metrics that find visibility and pre-eminence. The risk is that all that we will be doing is perpetuating the power of metrics, but such a knowing approach towards resistance could engineer opportunities to challenge the logic of the numbers and perhaps even to challenge the logic of the circulating measures of metric power and what it is that they appear to make possible. But of course, this would again potentially lead to an expansion of the reach of metrics, and question marks have already been raised about taking the steps of knowing our own data (Nafus and Sherman 2014; Neff 2013).

It is possible that the way forward is not to imagine new ways to resist, but to look at how people already respond to the metrics in their lives. Wendy Espeland's suggestion is that in order to understand responses and reactions to quantification we should look at the narratives and stories that are attached to metrics. Such stories, Espeland (2015: 74) claims, 'will help us understand how people who make and are governed by indicators make sense of them, understand the stakes of their simplification and resist them'. Therefore it is by looking at how metrics strip out the narratives about those being measured, and how those metrics are later narrated, that we might understand power and resistance as it unfolds in people's lives (see Chaps. 4 and 6 for more on this). It is possible that both power and resistance might well be located in the sense-making processes to which Espeland refers here. The difficulty will be in keeping a broad sense of the power of metrics if we are to immerse ourselves in the way that those metrics carry meanings for people. One means of response here might be to conclude that as narratives are stripped out through metricisation, as Espeland has argued, we need to be ready to re-narrate them and to re-tell the stories that get lost.

Having provided these reflections, I wouldn't like this book to simply be read as an attempt to bemoan numerical thinking in contemporary society. Rather, I'd hope that this book might be used as a potential resource for understanding, analysing, and responding to the pressures, anxieties, and stresses that are promoted by metric power and to collectively think through their affects (I continue with this in Chap. 6). This requires nuance and a mobile set of conceptual resources that enable us to think through the different scenarios within which metric power plays out. My hope is that the resources here can be used to think with and to promote an understanding of the variegated implications and mobilisations of metric power. This book is far from a complete resource, nor is it a final destination for the conceptual and empirical work that is required, but it might offer some scope for examining the power of metrics and for extending the dialogue around and understanding of the role of metrics in people's lives.

In all of this, there is an underlying sense that measuring the social and the human is also to shape and discipline the social and the human. As Espeland and Stevens (2008: 414) have put it, 'numbers can also exert discipline on those that they depict'. The result, as we have seen, is, as they add, that measures 'designed to describe behaviour can easily be used to judge and control it' (Espeland and Stevens 2008: 414). Measures are at the same time judgements, and therefore have the capacity to control the thing being measured (to try to tackle this problem more directly, I've added Chap. 6 as a coda to this book). In attempting to understand such a set of complex and interwoven phenomena, I have argued that we need to look at the consequences of the circulation of metrics. The outcomes of such processes are vast, incorporating as they do anything from the nation state, to corporations and organisations, as well as bodies, knowledge, culture, and so on. This is a multi-scalar problem and careful attention is needed to work across such scales. Perhaps, if we are to pick out one feature, the question of visibility is of particular importance. Metric power is based, as we have seen, on the capacity to make visible. If there is one characteristic of metric power that we might want to take as central, and if we were only to concentrate on one feature, it should be that of visibility and invisibility, attention and non-attention, recognition and non-recognition. It is in such divisions that metric power gains its leverage. Therefore, it is in these distinctions that we may begin to prize open the power that metrics hold whilst locating the other features that I have outlined in this conclusion and in the earlier chapters. The 'cold intimacies' (Illouz 2007) of contemporary capitalism are to be found in these distinctions and in the decisions concerning visibility and value that are afforded by metrics. If we look carefully, it is also in this space and in these metric-based visibilities that we will find the contestation between value and values to which Bev Skeggs (2014) is pointing us. Stuart Elden (2006: 175) has suggested that as 'a response to the notion of quantity becoming a quality in itself, this notion of existence beyond number is an important issue'. This does force us to ask, in the context of the kind of metric power I have described here, whether there is existence beyond number, and if there is, is it being chipped away at by the reach of those measures and calculations. The territory that both Elden and Skeggs outline, in their very different ways, is a space in which the reach of metric power is transforming the relation between value and values. These are the points of contest, a political set of boundaries and battle lines, the points of contact between what is measured and what is not.

The difficulty is that just because something is not measured and therefore has an existence beyond number does not mean that it escapes the logic of metric power, nor does it mean that it is empowered or that it can be a value beyond valuation. Rather, as we have seen in the earlier discussions, those things outside of the reach of metrics are not necessarily outside of the reach of metric power. As we have seen, not being counted can be to render invisible, marginal, and devalued. The power of metrics goes beyond the reach of the metrics themselves to have implications for all aspects of the social world, even those that are not (yet) measured. And let us not forget metric power's tendency to expand and spread into new areas, to measure new things (as we saw in Chaps. 1 and 2). As White (2014: 131) observes in relation to the importance of things which may not be measurable, 'in an increasingly quantitative world…exemplified by Big Data and virtually limitless computing power, the elegant simplicity of numbers threatens to crowd out more nuanced, qualitative concepts'. Metric power can comfortably *crowd out* alternative ways of thinking and hustle conceptions of what is worthwhile. This is where we need to think, very seriously, about its consequences.

With all of this in mind, what is pressing for social theory and social research more broadly is a more nuanced and detailed engagement with forms of measurement, the circulation of metrics, and what these circulating metrics make possible (or impossible). This is not an entirely new problem, as I have shown here, but it is a problem that needs to be reaffirmed to respond to the intensity and scope of contemporary systems of measurement and their potential reach through mediated dissemination. The conceptual frameworks at our disposal need to be reshaped to respond to these questions and transformations. This is now pressing when we reflect on the global development of 'variegated neoliberalization' (Brenner et al. 2010) and the escalation of data assemblages and systems of measurement. It is these very systems of measurement which, in turn, may be extending the span of neoliberal forms of governance. Based upon his deft analysis of the interweaving histories of sociology and neoliberalism, Nick Gane (2014b: 1103) warns us that 'neoliberalism is a complex and multi-faceted project and it would be a mistake to reduce it to a single epistemological position or commitment'. Nevertheless, in the context in which it is developing, in whatever form, the foundations of its machine will be found in systems of measurement and the circulation of metrics. Beyond this though, understanding systems of measurement and the circulation of their outputs is central to understanding the functioning and performances of the social and cultural world.

It is in systems of measurement that we will find the very *mechanisms of competition* (see Chap. 1). Systems of measurement are a central part of the way that the social is imagined and produced, they are technological but they are ushered in by a certain rhetoric and some attendant culturally determined desires. Measurements are afforded by infrastructural mechanisms, but these mechanisms are also discursive and cultural. Systems of measurement become powerful in shaping the rules of competition. What can possibly be measured becomes the logistical questions within which the games are played. This is the starting point, *with systems of measurement as the very mechanisms of competition*. But we need then to move beyond this to see how these measures circulate through the social world. This combination of metrics and their dissemination then, potentially, defines what is possible. Under the conditions of neoliberal governance, and the processes of neoliberalisation, it is this intersection to which our attention should be drawn. The production and dissemination of metrics is central to understanding the neoliberal 'art of governance' (Foucault 2008). It is here that the battle lines are drawn, where power dynamics are instantiated and where the rules become realities. As such, it is at the points where measurement and circulation meet that we will see how possibility is being founded. This is the terrain that needs further conceptual attention. My hope is that this book provides some ground from which such a conceptual set of resources might be explored. It is hoped that the concept of *metric power* might help to draw some analytical attention towards the telling and important intersections of measurement, circulation, and possibility. When trying to understand the role of metrics we need to keep this trilogy of components in mind. Beyond this though, and in more general terms, when thinking about new types of digital, big, by-product or transactional data, we should be thinking both genealogically about their origins whilst also thinking conceptually about the politics and power dynamics that reside not only in these data but in the ways that they are described, presented, and marketed to us. Metric power has not suddenly arrived, sneaking up behind us, and it is not purely material in its form. Metric power should be seen in its historical dimensions and it should be explored both as a material presence and as a central part of the way that the contemporary age is being imagined and reimagined in light of the fantastical possibilities that metrics are said to have. To study and unpick metric power is to think historically and materially whilst remaining attentive to the powerful flights of the imagination and the rhetoric of objective powers that come with the integration of systems of measurement into our lives. To do this, as I have argued here, we might begin with a focus on the interweaving of measurement, circulation, and possibility in both our everyday lives and in the organisational structures of various types within which those lives are lived and experienced.

References

Amoore, L. (2013). *The Politics of Possibility: Risk and Security Beyond Possibility.* Durham & London: Duke University Press.

Beer, D. (2015c, August 7). When 'special measures' become ordinary. *Open Democracy.* Accessed November 23, 2015, from https://www.opendemocracy.net/ourkingdom/david-beer/when-'special-measures'-become-ordinary

Bowker, G., & Star, S. L. (1999). *Sorting things out: Classification and its consequences.* Cambridge, MA: MIT Press.

Brenner, N., Peck, J., & Theodore, N. (2010). Variegated neoliberalization: Geographies, modalities, pathways. *Global Networks, 10*(2), 182–222.

Brown, W. (2015b). *Undoing the demos: Neoliberalism's stealth revolution.* New York: Zone Books.

Burrows, R., & Ellison, N. (2004). Sorting places out? Towards a social politics of neighbourhood informatization. *Information Communication and Society, 7*(3), 321–336.

Burrows, R., & Gane, N. (2006). Geodemographics, software and class. *Sociology, 40*(5), 793–812.

Dardot, P., & Laval, C. (2013). *The new way of the world: On neoliberal society.* London: Verso.

Davies, W. (2014). *The limits of neoliberalism.* London: Sage.

Day, R. E. (2014). *Indexing it all: The subject in the age of documentation, information, and data.* Cambridge, MA: MIT Press.

Desrosières, A. (1998). *The politics of numbers: A history of statistical reasoning.* Cambridge, MA: Harvard University Press.

Dodge, M., & Kitchin, R. (2009). Software, objects, and home space. *Environment and Planning A, 41*(6), 1344–1365.

Doria, L. (2013). *Calculating the human: Universal calculability in the age of quality assurance.* Basingstoke: Palgrave Macmillan.

Elden, S. (2006). *Speaking against number: Heidegger, language and the politics of calculation.* Edinburgh: Edinburgh University Press.

Espeland, W. N. (1997). Authority by the numbers: Porter on quantification, discretion, and the legitimation of expertise. *Law and Social Inquiry, 22*(4), 1107–1133.

Espeland, W. (2015). Narrating numbers. In R. Rottenburg, S. E. Merry, S. J. Park, & J. Mugler (Eds.), *The world of indicators: The making of governmental knowledge through quantification* (pp. 56–75). Cambridge: Cambridge University Press.

Espeland, W. N., & Stevens, M. L. (2008). A sociology of quantification. *European Journal of Sociology, 49*(3), 401–436.

Farlow, A. (2015). Financial indicators and the global financial crash. In R. Rottenburg, S. E. Merry, S. J. Park, & J. Mugler (Eds.), *The world of indicators: The making of governmental knowledge through quantification* (pp. 220–253). Cambridge: Cambridge University Press.

Foucault, M. (2002a). *The order of things.* London: Routledge.

Foucault, M. (2002b) *Power: Essential works of Foucault 1954–1984* (Vol. 3). London: Penguin.

Foucault, M. (2007). *Security, territory, population: Lectures at the Collège de France 1977–1978.* Basingstoke: Palgrave Macmillan.

Foucault, M. (2008). *The birth of biopolitics: Lectures at the Collège de France 1978–1979.* Basingstoke: Palgrave Macmillan.

Gane, N. (2012). The governmentalities of neoliberalism: Panopticism, post-panopticism and beyond. *Sociological Review, 60*(4), 611–634.

Gane, N. (2014b). Sociology and neoliberalism: A missing history. *Sociology, 48*(6), 1092–1106.

Gordon, C. (1991). Governmental rationality: An introduction. In G. Burchill, C. Gordon, & P. Miller (Eds.), *The Foucault effect* (pp. 1–51). Chicago: The University of Chicago Press.

Hacking, I. (1990). *The taming of chance.* Cambridge: Cambridge University Press.

Harvey, D. (2005). *A brief history of neoliberalism.* Oxford: Oxford University Press.

Illouz, E. (2007). *Cold intimacies: The making of emotional capitalism*. Cambridge: Polity Press.

Kennedy, H. & Moss, G. (2015). Known or knowing publics? Social media data mining and the question of public agency. *Big Data and Society*, 1–11. doi: 10.1177/2053951715611145.

MacKenzie, D. (2015, May 21). On 'Spoofing'. London Review of Books, p. 38.

McNay, L. (2009). Self as enterprise: Dilemmas of control and resistance in Foucault's *The Birth of Biopolitics*. *Theory Culture and Society, 26*(6), 55–77.

Miller, P., & Rose, N. (2008). *Governing the present*. Cambridge: Polity Press.

Mirowski, P. (2013). *Never let a serious crisis go to waste: How neoliberalism survived the financial meltdown*. London: Verso.

Nafus, D., & Sherman, J. (2014). This one does not go up to 11: The quantified self movement as an alternative big data practice. *International Journal of Communication, 8*, 1784–1794.

Neff, G. (2013). Why big data won't cure us. *Big Data, 1*(3), 117–123.

Porter, T. M. (1995). *Trust in numbers: The pursuit of objectivity in science and public life*. Princeton, NJ: Princeton University Press.

Porter, T. M. (2015). The flight of the indicator. In R. Rottenburg, S. E. Merry, S. J. Park, & J. Mugler (Eds.), *The world of indicators: The making of governmental knowledge through quantification* (pp. 34–55). Cambridge: Cambridge University Press.

Rose, N. (1991). Governing by numbers: Figuring out democracy. *Accounting Organization and Society, 16*(7), 673–692.

Savage, S. (2009). Against epochalism: An analysis of conceptions of change in British Sociology. *Cultural Sociology, 3*(2), 217–238.

Savage, M. (2010). *Identities and social change in Britain since 1940: The politics of method*. Oxford: Oxford University Press.

Scharff, C. (2015). The psychic life of neoliberalism: Mapping the contours of entrepreneurial subjectivity. *Theory, Culture and Society*. Online first. doi: 10.1177/0263276415590164.

Skeggs, B. (2014). Values beyond value? Is anything beyond the logic of capital? *The British Journal of Sociology, 65*(1), 1–20.

Tyler, I. (2015). Classificatory struggles: Class, culture and inequality in neoliberal times. *The Sociological Review, 63*(2), 493–511.

White, M. D. (2014). *The illusion of well-being: Economic policymaking based on respect and responsiveness*. New York: Palgrave Macmillan.

6

Coda: Metric Power and the Production of Uncertainty (How Does Metric Power Make Us Feel?)

I'm at my desk. I'm attempting, possibly in vain, to write a book that might be rated at least '3*' by my peers. That is to say I'm toiling over a piece of writing that I hope might be considered to be 'internationally excellent' in the eyes of both internal and external research assessment panels. The burden of excellence weighs heavy on my keyboard. The implications of this are both substantive and affective. The substantive implications of this research measurement are to be found in the words that I write as I labour at my desk. Thus knowledge potentially becomes a product of the way it is measured. But what I would like to focus on in this chapter is what I will call *affective measures*. This is a concept concerned with how *metric power* makes us feel. This is not a phenomenon limited to academic research, although it has now been acknowledged to be an increasingly prominent part of the ordering, experiences, and emotions of academic life (see Gill 2010; Burrows 2012; Hockey et al. 2014). It has been argued that academics have come to be increasingly governed and disciplined by metrics (Rushforth and de Rijcke 2015: 118). I wish to look far beyond my own practice and those of other academics. I will put my 'writing shame' and 'academic anxieties' (Probyn 2010) to one side; rather, I want to think more broadly about the interwoven relations

© The Editor(s) (if applicable) and The Author(s) 2016
D. Beer, *Metric Power*, DOI 10.1057/978-1-137-55649-3_6

between measurement and affect—and the affective implications of intensifying systems of measurement.

This chapter, or coda as I've called it, is about how systems of measurement make us feel and how they impact on our attitudes and approaches to the social world. This seemed like a necessary appendix given the need for us to think conceptually about the implications, embodiment, and experiences of metric power. This chapter is not necessarily focused on what is measured and how, but it is concerned with the way that we respond to those systems of measurement. Let me begin then with a claim that may be considered to be speculative and contentious, but its import means that I'm driven to make it anyway: most people are now exposed to intensified forms of measurement, and as such, most people, whatever they do, are living with and experiencing *affective measures*.

This chapter attempts to work through these key assumptions and to develop this notion of *affective measures*—and thus it begins to sketch out the affective responses that we have to *metric power*. To do this, it begins by returning to the connections between broader political formations and the role of measurement. It asks what part measurement performs in the affects of neoliberal governmentality and how this plays out in our everyday lives. The focus in this opening section is upon the relations between neoliberal competition, metrics, and, crucially, *the production of uncertainty*. It then brings these observations together with work on affect theory, and particularly Margaret Wetherell's (2014) notion of 'affective practice', in order to imagine how we might understand metric power through the lens of affect. The chapter then continues by providing some discussion of how measurement can be seen to be affective in its consequences and outcomes. It points towards a series of ways in which we might appreciate the physical and emotional experiences of measurement and, in so doing, carves out possible ways for us to see that it is not solely the measurement itself that matters but also the affective responses that the possibility of measurement creates for individuals.

Just to add a note of clarification before continuing, Nik Brown (2015a: 119) has written recently about the potentially performative and 'marshalling' role played by the scales and indexes used to 'metricise the emotional states of cancer patients'. Brown's important article discusses an attempt to try to measure affect—which in itself may shape behaviour and outcomes in the ways we've discussed earlier in this book. In reverse to this, what I'm concerned with in this chapter is not the measurement of affect itself, but with the affective power of the measures to which we are exposed.

Measurement and the Production of Uncertainty

We can begin then by returning to the connections between the broader political economy and systems of measurement, but in this case we can focus upon the way in which competition provokes senses of uncertainty. As we have discussed in detail in this book, measurement and calculation are clearly central to what is often referred to as neoliberalism or neoliberal governance (for a history of this connection, see Foucault 2008; and for a further historical illustration of the 'calculating character of modern times', see Simmel 2004: 481–483). We have already discussed this in some detail in Chap. 1 and throughout this book. The measurement and calculation of populations might have a long history, but as detailed in the previous chapters of this book, these measures have intensified as a result of growing data infrastructures and the cultures of neoliberal governance.

As discussed in Chap. 1, neoliberalism, which is by now a fairly familiar concept, is based upon market ideals and the promotion of competition (see Gane 2012). We noted in these discussions how measurement and calculation are required to make visible the victories and failures, and for markets of any type to become ranking-based realities. Neoliberalism and its ethos of competition have been discussed in a range of excellent and wide-ranging works (as discussed in Chap. 1), so I will not discuss this again here. Rather, the general ethos of neoliberalism operates as a kind of social and political backdrop to a more specific angle that I wish to develop as an opening to this particular chapter. The focus, in this opening section, is upon the connection between neoliberalism, competition, and what we might think of as *the production of uncertainty*.

In his key book on neoliberalism and competition, Will Davies (2014: xi) argues that uncertainty 'is a key concept for neoliberalism'.

Uncertainty, he suggests, is central to facilitating competition and entrepreneurialism—which, in turn, we might link to observations about broad shifts towards precarious labour, precarity, and insecurity (Gill and Pratt 2008). Davies (2014: xi) adds that in a 'specifically economic sense uncertainty is an effect of multiple, competing actors, operating according to various conflicting strategies in identifiable market places, established institutions and global arenas'. In other words, the layered competitions of neoliberal formations inevitably play on the uncertainty of those involved. A level of uncertainty is required to fuel competition. This, it would seem, has some parallels with Power's (2007) comparable notion of 'organized uncertainty'—with 'uncertainty at the centre of organizational design principles' (Power 2007: 7). As a result, competition 'imposes precarity and stress upon individuals' (Davies 2014: xi). These broad political arrangements then, and the rise of visible forms of competition and ranking, filter down to individuals in the form of stress, precariousness, insecurity, and an unshakable sense of uncertainty. The production of uncertainty is designed into these strategies and techniques. As Davies (2014: xii) puts it, 'individuals are trained and "nudged" to live with certain forms of economic uncertainty'.

At the centre of this is the escalation of the means by which individuals, groups, and organisations can be compared, as discussed in the previous chapters. This escalation is based on the emergence of advancing systems of measurement and the experts who profess to find value in data (see Davies 2014: 30). The aim then is to rank. The objective is to find the means by which competition can be realised and exercised, to find ways of making competitive advantages and disadvantages visible. The means by which differences are established becomes the means by which competition operates and how it is then felt by the individuals involved. Replacing judgement with metrics (Davies 2014: 16) is to build up a 'faith' in the numbers and to promote the sense that these are objective measures (see Porter 1995). This then has consequences for those being judged and ranked. Frequently, 'league tables' are used to give 'an empirical and technical form to the competitive market ideal' (Davies 2014: 44). For Gane (2012: 628), league tables are one key instance of 'the introduction of techniques of measurement or audit that enable the direct comparison of institutions through the construction of classifications'.

The league table is a simplified representation of the different data that is drawn upon to promote and extend forms of marketised competition and, despite any concerns we might have about the data that informs them, they often become realities. The very power of metrics is in their capacity to simplify, whilst also elaborating, complex social formations, as Espeland (2015: 56) has recently argued. Indeed, league tables and other such modes of presentation are used to 'manufacture marketized forms of competition where previously they did not exist' (Gane 2012: 632). These ideas though have been discussed earlier in this book, the question here is what this means in terms of the individual consequences of these neoliberal forms of competition.

The results are powerful. These 'competitive processes…preserve an element of uncertainty in social and economic life' (Davies 2014: 188). The aim, Davies (2014: 188) argues, is to 'produce quantitative facts about the current state of competitive reality, such that actors, firms or whole nations can be judged, compared and ranked'. Here uncertainty is coupled with being ranked and compared. We might add that, as I will explore, there are further complexities to how people are ranked and what type of uncertainty this evokes. We are often ranked from multiple perspectives using different combinations of data this means that, even if you do well in some of the measures, you are always likely to fail somewhere. There is always some leverage within the data and, therefore, some means of creating uncertainty. There is always an area where the data shows that you are behind, or where your position looks fragile in relation to your competition. So, this uncertainty might be unpicked further to reveal its various forms and how it manifests itself in response to the measures being used. Ros Gill, for instance, focusing upon the 'psychosocial aspects of neoliberalism', suggests that by focusing upon 'experience' we can reveal the 'costs' of neoliberalism which are frequently felt as 'insecurity, stress, anxiety and shame' (Gill 2010: 241). As Christian Scharff (2015) has similarly found in her study of the 'psychic life of neoliberalism', a number of her interviewees indicated the prominence of feelings of anxiety, doubt, and insecurity. These feelings were instigated by a sense of not knowing what was coming. Scharff's (2015: 10) interviewees—who had trained themselves to 'embrace risk', 'hide injuries', and 'compete with themselves'—revealed the scale to which 'insecurities provoked anxieties'. Such experiences might well reveal more of the operations of metric power, particularly as it comes to act upon individual bodies.

The connection here is likely to be between risk and uncertainty. Mirowski (2013: 121) has argued that 'the neoliberal celebration of risk is woven throughout everyday life in the modern era'. Neoliberalism demands risk taking and enjoys provoking the sense of being in constant risk, we might imagine that this is likely to be complemented by a sense of uncertainty about the future or the potential consequences of risk and risks. We are returned to the neoliberal self discussed in Chaps. 1 and 5, but this time we see the drivers. Mirowski (2013: 119) adds that 'participation in neoliberal life necessitates acting as an entrepreneur of the self: unreserved embrace of (this version of) risk is postulated to be the primary method of changing your identity to live life to the fullest'. The neoliberal outlook is to embrace risk and therefore to embrace uncertainty, these become the drivers for entrepreneurialism and individualised action. We can also turn here to Lilley and Lightfoot's (2013) work on the embodiment of neoliberalism for an account of how 'precarity' and 'competitive self-interest' are used to fashion entrepreneurship and 'entrepreneurial alertness'.

For Davies, there are two main types of uncertainty to focus upon here. The first is 'competitive uncertainty'. This is a form of 'uncertainty that arises within the arranged economic "game" as a result of multiple actors, all pursuing conflicting and distributed agendas' (Davies 2014: 149). Competitive uncertainty is the form of uncertainty that is deliberately cultivated to enable and provoke competition. As Davies (2014: 149) explains, competitive uncertainty 'is the form of uncertainty that neoliberals have always celebrated, not least because they claim to have the tools through which this uncertainty can be rendered periodically empirical, intelligible and manageable'. Neoliberal governance aims to produce this type of uncertainty and then uses methods and techniques of measurement to manage it. This contrasts with 'political uncertainty' which is a form of 'uncertainty that challenges the very terms on which doubt and judgment are to be performed' (Davies 2014: 149). In short, the first form of uncertainty is desired and is the driving force of neoliberal social control. The second form is an uncertainty in the core principles of neoliberal governance and as such casts its desirability into doubt. In this chapter, I am focusing solely on 'competitive uncertainty'. That is to say that I'm focusing upon the type of uncertainty that is deliberately produced by and through competition and measurement. This is the form of uncertainty produced as 'competitiveness evaluations and comparisons are produced in order to *frighten enthuse* and *differentiate*...designed to catalyse entrepreneurial competitive energies' (Davies 2014: 138). We are working then with Davies' crucial observation that 'the pragmatic purpose of competitive "scoreboards" is not to achieve a form of peaceful consensus...but to nurture existential anxieties' (Davies 2014: 138; see also Konings 2015: 94). These anxieties and uncertainties, this would suggest, are produced by the forms of measurement that facilitate neoliberal marketised competition. Or, as Gane (2012: 631) has put it, this is 'governmentality through surveillance to promote competition'. Metrics, in their role of facilitators of competition, are central to this production of uncertainty.

What this would suggest is that precarity and uncertainty are produced on the ground through the deployment of various systems of measurement. We have an interesting set of triangular relations emerging here: Competition requires measurement. Neoliberalism requires competition. Competition requires and produces uncertainty. Measurement produces uncertainty and affords competition. In short, the systems of measurement typical of the art of neoliberal governance produce uncertainty— they are deeply affective. Indeed, Lisa Blackman (2012: 191) has argued that affect theory should be used to think about the 'processes of subjectification within contemporary neoliberal forms of governmentality'. This is a pressing problem, both for understanding political governance and control through measures, and for enhancing our understanding of the embodiment of those political formations. In terms of the production of uncertainty, we might imagine that our sense of uncertainty is produced by particular objectified metrics or, perhaps, by our expectation of these metrics and their outcomes. According to Konings (2015: 30), a key limitation of Foucaultian approaches is that 'neoliberal subjects are depicted as having their anxieties mostly under wraps, their insecurities proficiently managed by the relevant agencies'. This is probably accurate, although there is some sense in these positions that people only just manage to have their anxiety in check (see also Hill 2015: 40–54). With the inducement of the level of anxiety required to achieve optimum activity and productivity, there is likely to be overspill. As such, we might want to think about measurements as objects, and as Ahmed (2004: 125) has proposed, 'anxiety becomes attached to particular objects, which come to life not as the cause of anxiety but as an effect of its travels'. If we wish to make measurements affective, then we may wish to see how they travel through our lives.

Making Measures Affective

We might deduce that the measurement of our practices and performances create 'affective atmospheres' (Anderson 2009) that hang over us and shape our feeling of time and space. James Ash (2010) has used the helpful notion of 'architectures of affect' to understand the affective properties of video games. Our starting point though might be to think of this as a concept that represents the broader architectures of everyday life. That is to say that we might begin by thinking of systems of measurement as being part of the 'architectures of affect' within which everyday life is conducted. Although it might be a stretch to imagine that the ordinary spaces that we occupy have affect designed into them in the same way as a video game, nonetheless it provides us with a vision that in some ways resonates with Burrows' (2012) accounts of metrics and everyday data assemblages. As Ahmed (2010: 35), in a general account of happiness and objects, proposes, 'objects are sticky because they are already attributed as being good or bad, as being the cause of happiness or unhappiness'. Measures and metrics are objects with such potential capacities. The difficulty is how we might go about understanding metrics as affective. The problem is where to start if we want to think about how measurement is felt, how it is embodied, and how it can be seen to be experienced emotionally. Clearly, this produces for us a set of questions and possibilities that stretch far beyond the capacity of this chapter alone. What I would like to do here then is to open up these questions by drawing upon work on affect theory and the sociology of emotions. This conceptual framework can provide us with the beginnings of a toolkit for analysing what I call here *affective measures*, and for understanding how *metric power* works through the production of uncertainty.

A key argument of the crucial paper on the sociology of quantification by Espeland and Stevens (2008) is that we should think of numbers as 'deeds'. In this sense, we should approach them as we do words. They go on to argue that metrics should be seen as 'acts of communication whose meaning and functions cannot be reduced to a narrow instrumentality and which depend deeply on "grammars" and "vocabularies" developed through use' (Espeland and Stevens 2008: 431). In short then, metrics might be seen as *deeds*, that is as active, loaded, and consequential forms of communication. Metrics are communication with purpose and intention. We need to appreciate then the subtleties of these numerical deeds. The act of measuring is active and can be thought of as a form of practice in which something is being actively done or achieved. This would suggest that there is a need to develop an account of affective measures that thinks in terms of metrics as deeds and which therefore captures them as practices aimed at provoking affect. With this in mind, the materials I pull upon here are particularly useful in helping us to think about how objects and systems are affective as they go about stimulating and provoking embodied emotional responses, especially where those objects are incorporated into practices and experiences. In this case, to narrow things a little further, I will take Margaret Wetherell's (2012) concept of 'affective practice' as my focus. Amongst the complexity and conflicting diversities of the work on affect, Wetherell's concept is particularly useful in that it enables us to illuminate affect as it emerges, exists, and plays out in practice. Thus we can use her work to see how metric power plays out in the practices of measurement and in terms of the affective properties of metrics. But let me very briefly outline what is at stake with affect theory before moving towards the specifics of affective practice.

With the turn to affect, Lisa Blackman (2012: 1) has suggested that 'rather than talk of bodies, we might instead talk of brain-body-world entanglements'. It is in such entanglements that we find the value of the concept of affect in its broadest sense. Untangling these would be crucial to understanding the affects of metrics. As Wetherell generally concurs, there is a 'subtle, relational, back and forth shuttling and interweaving going on at all levels of the body/brain/mind' (Wetherell 2012: 50). Or, as Seigworth and Gregg (2010: 2) add, 'affect is synonymous with force or forces of encounter'. This is to treat 'bodies as assemblages' and to consider 'that bodies are open, defined perhaps by their capacities to affect and be affected' (Blackman 2012: 1). Blackman and Venn (2010: 10) have similarly noted that 'this paradigm of co-enactment, co-emergence and co-evolution assumes from the outset that we are dealing with thoroughly entangled processes that require a different analytical and conceptual language to examine'. It is in this broad sense of what affect might mean that we can see the wide variety of ways in which this type of theoretical perspective might be deployed (for an overview of these various versions of affect see Seigworth and Gregg 2010: 5–8). According to Blackman (2012: 5), for instance, 'affect theory enacts and brings together a number of approaches to affect which differ in the place they accord the "human" within their analyses'. The variety in this theoretical field comes then with the balance or prioritisation of the presence of the human in the mixture of competing social factors.

Despite these various cocktails of the human and non-human, the material and immaterial, in these versions of affect theory there are some commonalities (for critical accounts of these positions see Leys 2011; Wetherell 2014). Michael Hardt (2007: ix) says that a 'focus on affects certainly does draw attention to the body and emotions, but…[t]he challenge of the perspective of the affects resides primarily in the synthesis it requires'. As such, the perspectives might vary but the general problem that is encountered concerns the synthesis, the interweaving and integration of the social with bodies and emotions. Whatever tack is taken, this synthesis remains the problem and, in fact, it is the approach towards this synthesis that leads to disagreements and incompatibility in the variegated uses of the concept of affect. This type of synthesis of the 'entanglements' of the 'brain-body-world' to which Blackman refers has led Patricia Clough (2010: 224), a key figure in the 'affective turn', to talk about the need for an 'empiricism of sensation'. Clough (2008) is drawn towards notions of the 'biomediated body' and the body 'being informational' (Clough 2008: 9). For Clough, the changing media through which we experience the world is deeply affective and needs to be regarded as such. As Clough (2008: 2) argues, 'the biomediated body exposes how digital technologies such as biomedia and new media, attach to and expand the informational substrate of bodily matter generally'. This gives an early point of departure for our concept of affective measures, because we can comfortably see how our bodies become biomediated by such systems. This, it should be noted, is a similar point to that being made by those with an interest in biometrics (see Chap. 2). In a message of warning though, Ian Burkitt has argued that it is important that we do not see affect and emotion as somehow existing in their own right or outside of social relations, this, for him, would be a mistake. Burkitt's (2014: 12) claim is that affect 'is not a mystical force or a charge akin to an electric current, but is a material process of its own kind created by body-selves acting in relational concert'. As such, understanding the rhythms and movements of this concert becomes important. We are also reminded by Burkitt not to look for affect outside of social relations or as emerging out of anything but social connections. We might begin to see then how metrics are integrated in this concert of social relations and the generation of emotional responses.

We can begin then with this very broad take on what is at stake with the concept of affect. As this brief outline already hints, the possibilities for the concept are wide-ranging and the positions taken are not necessarily analogous. For this reason, we need to tread with care and try to be precise in the deployment of such a concept. One of the most productive means by which the concept of affect might be used—both in terms of the remit and aims of this chapter and more generally—is Wetherell's (2012, 2014) concept of 'affective practice'. This variation on affect serves to outline a more direct way in which affect might be used to analyse social relations, whilst also giving us a means, as I apply it in this article, for making measures affective. That is to say that it can be used to locate metrics within practices and to also see these integrations as being affective in their outcomes. Treating metrics as deeds is to see them as practices. More than this, it is to see them as deeds or practices that have affects.

Despite the type of chaos, messiness, and vitality that is often associated with accounts of affect, Wetherell (2012: 4) develops an account of affect that is interested in the ordering and patterning of social life. It is this interest in pattern and order that forms the rationale for the concept of 'affective practice' (see Wetherell 2012: 11–12). Wetherell is developing a version of the concept of affect that focuses upon its part in social order and the maintenance of patterns as well as in their disruption.

This does not mean that Wetherell evokes a more sedentary, passive, and immobile version of affect; rather, these contestations and fluxes are seen to be part of the ordering and patterning in the social world (Wetherell 2012: 24). We are looking then at the very components or stuff of social life, in close-up detail.

In broad terms, Wetherell's concept of:

> affective practice focuses on the emotional as it appears in social life and tries to follow what participants do. It finds shifting, flexible and often over-determined figurations rather than simple lines of causation, character types and neat emotion categories. (Wetherell 2012: 4)

Despite its focus upon order and pattern, the concept of 'affective practice' is still intended to draw attention to the emotional and to see social relations as transient and complex. It resists simple causation and even reductive categories, but yet maintains the importance of ordering and ordering processes. In other words, it is interested in the interrelations between affect and power. This then is an important conceptual development, for it deals with social ordering but it does so in the context of social relations, embodiment, and emotions. It is about seeing the ordering and patterning of something as personal and as potentially chaotic as emotions and bodily sensation, allowing us to potentially see how metrics order and shape these emotions and bodily sensations. As Wetherell (2012: 11) poignantly asks 'how can we engage with phenomena that can be read simultaneously as somatic, neural, subjective, historical, social and personal?'. In short, and crucially, Wetherell's concept is about understanding 'affect in action' (Wetherell 2012: 4). Again we see how it might be useful in assessing the affective properties of the processes involved in being measured and in the metric-based production of uncertainty.

Wetherell suggests that affective practice draws us to 'three lines of approach that need to be at the heart of new work on affect' (Wetherell 2012: 11). Let us briefly reflect upon each. First, Wetherell claims that affective practices are flowing. That is to say that they have a temporality, rhythm, and sequence in our lives. They have a 'duration'. As Wetherell (2012: 12) explains, 'affective practices unfurl, become organised, and effloresce with particular rhythms…the chronological patterning of these figurations, along with their sequencing and "parsing", is crucial'. Individual experience is a product of the flows that are afforded by our social circumstances, and these affective flows operate across a range of scales. According to Wetherell (2012: 13), 'affective flows can become articulated with large-scale social changes such as patterns of modernisation, rural-urban shifts, equality movements and the logics of capitalism'. As broader patterns transform or remain consistent, these flows then filter into the affective patterns of everyday experience. As such, affective practice should be seen as being defined temporally by broader flows and movements whilst also being applicable on a range of scales. The result, for Wetherell (2012: 13), is that 'affective practice is continually dynamic with the potential to move in multiple and divergent directions' and therefore 'accounts of affect will need to wrestle with this mobility' (Wetherell 2012: 13). Thus, we are faced with forms of affect that are mobile and transient but which are rooted in practice and experience. They have flows and rhythms.

Second, as we have already hinted, Wetherell claims that affective practice should be seen as being patterned. She argues that affect displays 'strong pushes for pattern as well as signalling trouble and disturbance in existing patterns' (Wetherell 2012: 13). Affect then is not just about disruption but about consistencies and repetition; it is about pattern as well as disturbance. These mobile patterns intermingle and overlay (Wetherell 2012: 13–14). Whether or not this is a fair assessment, a central issue that Wetherell has with affect theory is that it has tended to focus on disturbance and to neglect or even actively disregard pattern. Instead, Wetherell develops a vision in which bodies pattern into their social environments and take on their flows, movements, and structures. We might wonder how metrics allow for this incorporation and patterning of bodies, as they are counted, judged, and ranked. It is in these patterns, it would seem, that we can find the recursive feedback loops that we experience and which shape affective practices—again this might be seen to add an affective dimension to the discussions of recursivity and circulation in Chap. 3. The patterns Wetherell refers to here are not fixed, but are changeable and mesh into one another.

As a consequence of this patterning, we become knitted into our environment by affective connections and relations. Wetherell (2012: 14) suggests that we are 'very densely knotted in with connected social practices where the degree of knitting reinforces the affect and can make it resistant and durable, sometimes unbearably so'. Yet Wetherell also warns that 'attempts to find order can break down as the dynamism of the phenomenon, the fuzziness, and instability of any descriptions of affective states, and sheer exuberant and excessive possibilities of the body become apparent' (Wetherell 2012: 16). The balance here is in understanding both patterning and disruption. It is also to explore how those patterns mesh into our everyday lives, becoming irresistible and inescapable. Again, we can see how metrics might have the potential to impose patterns on our emotional lives, but we might also reflect on how these patterns come to have a defining and immovable presence.

Third and finally, Wetherell suggests that a focus on power, value, and capital is productive in understanding affective practice. We are returned here again to issues of scale (see Chap. 5). We have seen how Wetherell places affect within broader structural flows and patterns, the result is that affect can then be understood as part of power structures and systems of value generation. Affective practices can vary in terms of the scales in which they are located. But perhaps the most striking and telling point in this regard is Wetherell's argument that 'power works through affect, and affect emerges in power' (Wetherell 2012: 16). It is perhaps this point that is most significant in terms of understanding the affective capacities of systems of measurement. For Wetherell, power can operate via affect. That is to argue that systems of power can draw upon affect whilst affect is also a product of those power structures. This is a vital observation, it is to suggest that affect is part of the structures of power to which we are exposed and that structures of power are then also used to draw out affective responses. This gives affect an important recombinant place in understanding how power might operate through emotions and bodily sensation and how interventions might be made into their entanglements. Wetherell adds that 'power, then, is crucial to the agenda of affect studies' because 'it leads to investigations of the unevenness of affective practices' (Wetherell 2012: 17). In other words, understanding power through affect leads us to see how affective practices of certain types are unevenly distributed, and thus so is their power. This is to consider how systems and formations of power lead certain people to feel certain things more acutely than others. Ian Burkitt (2014: 156), drawing on Wetherell's work, similarly concludes that social 'relations are also power relations, and we cannot…separate people's emotional responses and judgments of value from the power relations in which they are located'. Burkitt's position, like Wetherell's, is that emotion is a product of social relations and, therefore, is a product of the power dynamics of those relations. When reflecting on power we need also to reflect on the role of affect in facilitating that power and its distributed implications and consequences. For the moment though, we can approach the practice of measuring as potentially being an affective practice. To be measured is to be exposed to a type of affective practice.

Affective Measures in Action

If we bring together the materials we have covered so far, then we can begin to explore how neoliberal forms of governance, based upon markets and competition, aim at producing uncertainty. Producing uncertainty, as a visceral and emotional response, requires various types of, to use Wetherell's term, 'affective practices'. I'd like to argue that metrics, if we think of them as a process of providing measures rather than as fixed objects, represent a type of affective practice that is geared towards the production of uncertainty. Metrics, as deeds, are a part and product of a series of affective practices that have the power to make people feel uncertain, precarious, and anxious. Thus we have *affective measures in action*, a practice designed to provoke the uncertainty required by neoliberal forms of competition. Systems of measurement afford these marketised competitions and thus are the means by which uncertainty is produced. So, I'd like to begin to move towards a conclusion by reiterating the suggestion that systems of measurement are deeply affective. The power held by what I have called metric power is, predominantly, in how it makes us feel. It is through their affective capacities that metrics can be used to promote or produce actions, behaviours, and pre-emptive responses. Indeed, if we apply the work of Wetherell to metrics, then this might be an area in which we can see power operating through affect. As Wetherell points out, the difficult part is to try to capture this affect in action and to work towards an understanding of affective practice. Neoliberalisation requires affective practices in order to produce the uncertainty required to fuel competition—affective measures are used to this end. So here I will just begin, in tentative and provisional terms, to explore how such a project might be developed and how we might think of the use of *affective measures* in the deployment and realisation of *metric power*.

When considering measurement as affective practice we should start with the industry of data analytics that has emerged in recent years (see also Beer 2015b). It is this industry that points to the potential of systems of measurement and how they are embedding themselves in organisations, lives, and practices. It is an industry that is illustrative of how we are being measured or, beyond this, how we might be measured. There is now a significant industry based around the use of big data in human resources and performance measurement, with companies offering advice, information, and services for 'Talent measurement' and 'Talent analytics' or for 'harnessing big data to boost employee performance' (for just one example of this type of service and rhetoric, see www.plushr.com; see also Deloitte 2015). Indeed, there are even industry competitions for the best and most innovative means of using data to manage performance (for links to these see Beer 2015c). This is an industry that provides the expertise to enable people and their performance to be judged through data. The proliferation of data has led to a vast and growing industry of data analytics solutions companies and consultants. As organisations have found the data about their products, services, customers, employees have escalated, they often lack the means or the technologies to get anything from the data (unless they have in house analytics). As a result, an industry has emerged, populated with experts and analysts, that provides the means by which organisations can find out what their data is telling them. The consequence is that performance and practice are measured in vast ways, often drawing upon what Louise Amoore (2011) has called 'data derivatives'—these are data sources that are used for purposes to which they were not necessarily intended. The result of the availability of data and the expert knowledge and techniques required to analyse it is emblematic of Mirowski's observations that 'a kind of "folk" or "everyday" neoliberalism has sunk so deeply into the cultural unconscious' (Mirowski 2013: 89). In these circumstances, data analytics solutions, in the form of software packages or analytics services, are absorbed into the affective practices of organisations and then become part of metric informed rankings and judgements. The *practice of measuring practice* becomes normalised and is even expected to be part of the structures and self-presentations of those perceived or self-consciously presenting themselves as forward-thinking organisations, that is, organisations that see themselves as fit to compete in the contemporary world (for an example of this, see Beer 2015b).

Finding the means by which the data assemblage can be used to monitor employees as well as customers is seen to be important in human resource management (see e.g. the exploration of the use of metrics to measure human capital and drive performance in Huus 2015). As this excerpt from an article on the use of metrics in human resource management suggests: 'Applying Big Data analytics to your employees' performance helps you identify and acknowledge not only the top performers, but the struggling or unhappy workers, as well' (Fallon 2014). In other words, metrics enable competition to be more firmly established within organisations and for employee practice and performance to be gauged in new and affective ways. Here we are both using measures to provoke affective responses, creating a sense of uncertainty to motivate, whilst also measuring the affects that those performance measures have (by judging the unhappiness of the worker and so on). As with the broader rhetoric around big data, the potential of new performance metrics is described as being transformative. The rhetoric suggests that new means are now available for exposing or extracting value from people. This is represented as being a shift towards a more objective and denser numbers-based system of measurement of performance—this is a variation on what Sara Ahmed (2004) has called 'affective economies'. As this illustrative passage from an industry source indicates: 'No matter what solution you choose, an analytics program moves you away from the traditional manual "reporting" process of performance measurement, helping make your staff more efficient, motivated and engaged' (Fallon 2014). The material change described here comes in the form of a move away from 'manual' or qualitative forms of 'reporting' towards metric-based systems (see also Huus 2015: 8; Deloitte 2015). This passage shows that these metrics are designed to be affective measures. That is to say that these practices of performance measurement are aimed at provoking and capturing affect, emotion, feelings, and body–mind entanglements (as discussed earlier in the chapter in relation to the work of Blackman 2012). Both of these excerpts from the performance management industry use measurement to deal with feelings and to provoke what is described as a positive emotional response. The first passage promotes the use of metrics to help the employer to deal with those who are unhappy or struggling (for further examples of this see Davies 2015a). The second passage promotes the use of metrics to motivate and engage. The more positive aspects of affective measures are drawn out in the rhetoric, but what of the other feelings they produce? This, of course, is only a snapshot, yet it is intended to be suggestive rather than comprehensive, to open up the questions rather than close them down. It should be noted though that there are plenty of other illustrative examples that could have been explored to make similar points (for a number of other illustrations see, for instance, Huus 2015).

This type of perspective on the rise of data analysis in the management of staff is echoed by John Bersin (2013), a human resource (HR) analyst, who wrote the following in an article for *Forbes* magazine:

> Companies are loaded with employee, HR, and performance data. For the last 30 years we have captured demographic information, performance information, educational history, job location, and many other factors about our employees. Are we using this data scientifically to make people decisions? Not yet. This, to me, is the single biggest Big Data opportunity in business. If we can apply science to improving the selection, management, and alignment of people, the returns can be tremendous.

Here we see a HR analyst proposing that the escalation of systems of measurement is a necessity. This is part of a series of examples of business analysts promoting the use of big data and measurement to inform decisions about hiring and to create new revelations for measuring staff performance or 'talent'. These are only a handful of very cursory observations, but they begin to provide some suggestion of the affective properties or dimensions of metric power.

Using the three strands that Wetherell suggests are central to her concept of 'affective practice' can assist in developing an understanding of how affective measures operate. First, the flow of measurement is clearly temporally defined and is often rhythmic in its implications. There are periods of measurement in which the emotions are ramped up to, as Wetherell notes, potentially 'unbearable levels' as our bodies knit with the temporality of the systems of measurement. We have annual reviews, periods of special measures, periodic reviews, assessment periods, the publication of results and league tables, and the like. But, we also have, in some instances, ongoing, real time, and potentially always-there measurements of our practices. Those working in a call centre are constantly monitored and their practices measured in real time by the systems they are connected into as well as through periodic reviews. In other industries, global positioning systems monitor movements, such as with supermarket or parcel deliveries. And, of course, we always have our social media profiles and smartphones and the like. The potential to be measured is a constant presence in our lives as a result of the connected environments in which we live. So, data infrastructures mean that the temporality of measurement can vary; it can be quick or slow, coming in fits and starts—even if it is actually always residing there, constantly in the background. There are moments and periods where measures are felt with greater force.

I would also add though that even the sense of foreboding that can come with distant moments of measurement being on the horizon, means that despite potential ebbs and flows affective measures are always there, nagging at us. There are though, those visible moments when some feedback arrives in the form of questionnaire results or when a crucial assessment or set of results comes in, or when competitor performances or rankings are published so that we see ourselves in relation to the performance of others, or when a new form of data gets visualised to allow us to see what we are perceived to be doing right or wrong, and so on. These are, of course, peaks in affective measures, and the consequences are very real for those who are being measured. But the power of these types of metrics is that they are either always there or on the horizon. We have then an interplay of the always-there and the momentary escalation to consider when thinking about the affective presence of metric power in our lives. Considering the temporality of the physical and emotional responses systems of measurement provoke is crucial to fully understanding the relations between metrics and power.

Affective measures then are temporally defined and are rhythmic. They may always be there, in the background, but they lead to moments of heightened experience. The affective capacities of systems of measurement draw us towards a sense of being measured, of having been measured, and of the material outcomes and consequences of those measures. We are stuck in measurement cycles, whilst also being constantly exposed to potential measurement. We need only look at the way that school teachers speak of the 'anxiety', the 'gut-wrenching fear', the lack of sleep, and so on when describing the experience of anticipating the periodic visit of school assessors or the receipt of the students' results (e.g. read Anonymous 2014) to see how obviously affective the flow of measurement is—and this is before the results are revealed; it is the anticipation of measurement that is so affective. Similarly, we can look at recent accounts of the data heavy new media working environments to see how metrics are used to reportedly generate anxiety in workers, with 'anxiety-provoking' meetings scheduled to reflect on the individual's metrics (Kantor and Streitfeld 2015: 11).

Clearly then the results of systems of measurement are about patterns. That is to say that they are designed to show how you performed against your competitors (be they the people you work with or others who are competing for the same business as you). Rankings, league tables, charts, tables, and visualisations are all material instantiations of the attempt to find patterns and to use them to provoke action. Metrics are designed to find patterns so that these can be used in what is considered to be strategic thinking. Systems of measurement are used to find patterns so that attempts can be made to either reinforce or break them. The visualisations that are often used to analyse data and metrics are themselves social patterns but, more importantly, they are used to understand and to respond to social patterning. In other words, they provoke responses and actions. Systems of measurement enable patterns to be visualised so that it can be decided if they are desirable patterns or not. As new data forms are analysed and envisioned in new ways, so the practices of organisations change with implications for those working in them or in the sector. New means are found for measurement and their dissemination to be affective or to provoke new feelings.

Finally, thinking of affective measures through Wetherell's concept of affective practice leaves us in no doubt of the power dynamics that are at play. As Gill and Pratt (2008: 21) argue, affect is not just about resistance it can also 'bind us into capital'. Affective measures are all about value—finding value, forcing value, capturing value, and attempting to locate value in people and practices. As we have discussed throughout this book, systems of measurement are the mechanisms of the neoliberal agenda and the forms of competition that are required by it. We can now give these an emotional and bodily presence. The more the potential for being measured, the more affective these systems are and the greater their capacity for affording variegated competition to emerge. People will feel affective measures unevenly, and their consequences will not be distributed in equitable ways. We need only look back across the previous chapters to see metrics as being directly linked to power, capital, and value. Here we see how their deployment connects the body and our emotions directly into these broader power structures.

Some Closing Reflections on Affective Measures

Of course, this chapter is highly provisional in its aims. It is included in this book as a bridge into further work that might explore the feelings generated by and through metric power. This chapter is intended to be suggestive of how we might expand upon and explore the everyday appropriation and experiences of metrics whilst maintaining a broader sense of the political formations, structures, and cultures of which they are a part. As such, this chapter represents an attempt to begin to think about where we might go in order to expand our understanding of metric power. One key question concerns the way that metrics make us feel. My argument is that the power of metrics is in exactly this; it is in how those metrics make us feel before, after, or whilst we are exposed to those systems of measurement—which in some cases will be continual. The point here is that affective measures are designed to provoke the uncertainty that is typical of the cultivation and spread of forms of neoliberal competition. Metrics are used to manufacture uncertainty and to drive entrepreneurialism and self-training. This is the argument that we have begun to unravel in this chapter, but which I would suggest now needs much more detailed attention. To return to the preface of this book, this is where the joining of the dots becomes a little fainter and where the invitation for the reader to join in becomes more pronounced, for the moment at least.

With this in mind, and to conclude, we can say that uncertainty is produced by the politics of social relations instantiated in *the measure*. Systems of measurement are often highly individualised and individualising. This is part of how they operate affectively. They target, cajole, and provoke. They are aimed at stimulating anticipation and uncertainty—often coupling these with senses of insecurity and precarity. Affective measures then can be understood to be a crucial and central part of what Brenner et al. (2010) refer to as the ongoing processes of 'variegated neoliberalization'. Neoliberal governance, it would seem, is about the pursuit, maintenance, and production of uncertainty, particularly as audits are 'internalised' (Gill 2010: 235). The production of uncertainty drives and facilitates competition, whilst competition then produces uncertainty. Metrics become the means and mechanism for the reproduction of these recursive relations. Metrics, in short, can be used to produce uncertainty. It is this uncertainty that drives energies towards entrepreneurialism and provokes the desire for competitive advantage (or the avoidance of failure). Systems of measurement and the dissemination of metrics facilitate this uncertainty, and as such, it is the affective properties of measurement, the way they make us feel, that is so powerful in the art of neoliberal governance.

Measurement and calculation are often linked to neoliberal governance but we might take this a step further to think not just about these forms of measurement, but how they are experienced by individuals. Because these are *affective measures*, they lead individuals to self-monitor, to pre-empt the systems, to play the game, to act before being measured. Affective measures do not simply act as posterior mechanisms for improvement and performance management; they are also a pre-emptive or a priori presence in the lives of individuals. We know we are being measured. We may not be sure how or to what end, but the fact that we know makes these measures affective.

What makes these systems of measurement so powerful is the affective responses that they provoke, not just merely the fact that they exist. This is an important step to take in understanding how these wider political structures play out in the lives of individuals. Affective measures are powerful because they operate by performing both *posterior* and a priori checks on peoples' lives, behaviours, practices, and actions. The power of affective measures is in how we anticipate them and how they make us feel. The affective responses that these forms of measurement create are what give powerful leverage to those who are in a position to exercise it. The concept of *affective measures* works to give an emotional and bodily dimension to the broader concept of *metric power* that I have developed in this book.

In Chap. 3, we briefly discussed Espeland and Sauder's use of the concept of 'reactivity' to understand how people respond to the meanings they associate with measures and rankings. As we saw, they argued that we need to understand what they call the 'mechanisms' and 'effects' of that reactivity. Elsewhere, Espeland and Stevens (2008: 412) also argue that 'measurement intervenes in the social world it depicts'. They add that 'measures are reactive; they cause people to think and act differently' (Espeland and Stevens 2008: 412; see also Beer 2015b). Reactivity is used in Espeland's co-authored work to understand the implications of social measures for agency, especially in terms of sense making and interpretation. The point is that in order to further understand reactivity to measures, we need to grasp the 'mechanisms of reactivity'. Espeland and Sauder's (2007: 10; italics in the original) contention is that 'explaining *how* rankings are reactive, *how* they produce the changes that they produce, helps us better understand why these measures create such powerful effects'. My point here is that a notion of affective measures and an understanding of the production of uncertainty is a key way to understand the mechanisms of reactivity to which Espeland and Sauder refer. It is by understanding the emotional and corporeal responses that are provoked by metrics that we can understand what drives the reactions that people have to them. In other words, the visceral feelings that measures create in us can be seen as an instigating factor in reactivity. A key instigating factor in how measures produce outcomes, behaviours, and practices is in how they make us feel.

The key point that I would like to close with is that the power of systems of measurement is often not directly in what they track, capture, or allow us to compare, but rather it is in how the possible outcomes of being measured make us feel—and thus twist and cajole what it is that we do. Our bodies become knitted into these systems of measurement, we do not experience them consistently or evenly. The anticipation, the expectation, the worry, the concern, the fear of failure, the insecurity that comes with potential visibility, and so on, are all very powerful. Measurement works as a system of governance, and self-governance, because of its affective capacity. We are affected by how the metrics are likely to treat us, by what they make visible, and then again by how we are ultimately treated by them. When thinking of the social place and implications of metrics—as a part of the competition central to neoliberal processes—we should be thinking of affective measures. It is here in this type of conceptualisation that we might get at the subjectivity of measurement and our emotional and bodily responses to metrics. The concept of affective measures sensitises us to these bodily and emotional experiences of measures. However accurate/inaccurate or representative/misrepresentative they might be, measures are nearly always affective as they structure and order our practices and allow us to be compared and judged. *Metric power* works through the uncertainty it produces. The way to get a critical understanding of measurement, and the way to make a positive intervention, is to see metrics as being a deeply affective feature of wider systems of politics and governance. One important question that I think we need to ask in order to further develop notions of metric power simply concerns the way that metrics make us feel.

References

Ahmed, S. (2004). Affective economies. *Social Text, 22*(2), 117–139.
Ahmed, S. (2010). Happy objects. In M. Gregg & G. J. Seigworth (Eds.), *The affect theory reader* (pp. 29–51). Durham, NC: Duke University Press.

Amoore, L. (2011). Data derivatives: On the emergence of a security risk calculus for our times. *Theory Culture and Society, 28*(6), 24–43.

Anderson, B. (2009). Affective atmospheres. *Emotion, Space and Society, 2*(1), 77–81.

Anonymous. (2014, May 24). Secret teacher: Why are we really put through the pain of ofsted inspections? *Guardian.* Accessed November 12, 2014, from http://www.theguardian.com/teacher-network/teacher-blog/2014/may/24/secret-teacher-ofsted-inspections-education

Ash, J. (2010). Architectures of affect: Anticipating and manipulating the event in processes of videogame design and testing. *Environment and Planning D: Society and Space, 28*(4), 653–671.

Beer, D. (2015b). Productive measures: Culture and measurement in the context of everyday neoliberalism. *Big Data and Society, 2*(1), 1–12.

Beer, D. (2015c, August 7). When 'special measures' become ordinary. *Open Democracy.* Accessed November 23, 2015, from https://www.opendemocracy.net/ourkingdom/david-beer/when-'special-measures'-become-ordinary

Bersin, J. (2013, February 17). Big data in human resources: Talent analytics comes of age. *Forbes.* Accessed November 12, 2014, from http://www.forbes.com/sites/joshbersin/2013/02/17/bigdata-in-human-resources-talent-analytics-comes-of-age/

Blackman, L. (2012). *Immaterial bodies: Affect, embodiment, mediation.* London: Sage.

Blackman, L., & Venn, C. (2010). Affect. *Body and Society, 16*(1), 7–28.

Brenner, N., Peck, J., & Theodore, N. (2010). Variegated neoliberalization: Geographies, modalities, pathways. *Global Networks, 10*(2), 182–222.

Brown, N. (2015a). Metrics of hope: Disciplining affect in oncology. *Health, 19*(2), 119–136.

Burkitt, I. (2014). *Emotions and social relations.* London: Sage.

Burrows, R. (2012). Living with the h-index? Metric assemblage in the contemporary academy. *Sociological Review, 60*(2), 355–372.

Clough, P. T. (2008). The affective turn: Political economy, biomedia and bodies. *Theory Culture and Society, 25*(1), 1–22.

Clough, P. T. (2010). Afterword: The future of affect. *Body and Society, 16*(1), 222–230.

Davies, W. (2014). *The limits of neoliberalism.* London: Sage.

Davies, W. (2015a). *The Happiness Industry: How the Government and Big Business Sold Us Well-being.* London: Verso.

Deloitte. (2015). *Global human capital trends 2015: Leading in the new world of work.* London: Deloitte University Press.

Espeland, W. (2015). Narrating numbers. In R. Rottenburg, S. E. Merry, S. J. Park, & J. Mugler (Eds.), *The world of indicators: The making of governmental knowledge through quantification* (pp. 56–75). Cambridge: Cambridge University Press.

Espeland, W. N., & Sauder, M. (2007). Rankings and reactivity: How public measures recreate social worlds. *American Journal of Sociology, 113*(1), 1–40.

Espeland, W. N., & Stevens, M. L. (2008). A sociology of quantification. *European Journal of Sociology, 49*(3), 401–436.

Fallon, N. (2014, September 9). Big data: It's not just for customer insights. *Business News Daily.* Accessed October 27, 2014, from http://www.business-newsdaily.com/7099-big-data-employee-engagement.html

Foucault, M. (2008). *The birth of biopolitics: Lectures at the Collège de France 1978–1979.* Basingstoke: Palgrave Macmillan.

Gane, N. (2012). The governmentalities of neoliberalism: Panopticism, post-panopticism and beyond. *Sociological Review, 60*(4), 611–634.

Gill, R. (2010). Breaking the silence: The hidden injuries of the neoliberal university. In R. Ryan-Flood & R. Gill (Eds.), *Secrecy and silence in the research process: Feminist reflections* (pp. 228–244). London: Routledge.

Gill, R., & Pratt, A. (2008). In the social factory? Immaterial labour, precariousness and cultural work. *Theory Culture and Society, 25*(7–8), 1–30.

Hardt, M. (2007). Foreword: What affects are good for. In P. T. Clough & J. Halley (Eds.), *The affective turn: Theorizing the social.* Durham, NC: Duke University Press.

Hill, D. W. (2015). *The pathology of communicative capitalism.* Basingstoke: Palgrave Macmillan.

Hockey, J., James, A., & Smart, C. (2014). Introduction. In C. Smart, J. Hockey, & A. James (Eds.), *The craft of knowledge: Experiences of living with data* (pp. 1–18). Basingstoke: Palgrave Macmillan.

Huus, T. (2015). *People data: How to use and apply human capital metrics in your company.* Basingstoke: Palgrave Macmillan.

Kantor, J., & Streitfeld, D. (2015, August 15). Inside Amazon: Wrestling big ideas in a bruising workplace. *The New York Times.* Accessed August 19, 2015, from http://www.nytimes.com/2015/08/16/technology/inside-amazon-wrestling-big-ideas-in-a-bruising-workplace.html?_r=0

Konings, M. (2015). *The emotional logic of capitalism: What progressives have missed.* Stanford, CA: Stanford University Press.

Leys, R. (2011). The turn to affect: A critique. *Critical Inquiry, 37*(3), 434–472.

Lilley, S., & Lightfoot, G. (2013). The embodiment of neoliberalism: Exploring the roots and limits of the calculation of arbitrage in the entrepreneurial function. *The Sociological Review, 62*(1), 68–89.

Mirowski, P. (2013). *Never let a serious crisis go to waste: How neoliberalism survived the financial meltdown*. London: Verso.

Porter, T. M. (1995). *Trust in numbers: The pursuit of objectivity in science and public life*. Princeton, NJ: Princeton University Press.

Power, M. (2007). *Organized uncertainty: Designing a world of risk management*. Oxford: Oxford University Press.

Probyn, E. (2010). Writing shame. In M. Gregg & G. J. Seigworth (Eds.), *The affect theory reader* (pp. 71–90). Durham, NC: Duke University Press.

Rushforth, A., & de Rijcke, S. (2015). Accounting for impact? The journal impact factor and the making of biomedical research in the Netherlands. *Minerva, 53*(2), 117–139.

Scharff, C. (2015). The psychic life of neoliberalism: Mapping the contours of entrepreneurial subjectivity. *Theory, Culture and Society.* Online first. doi: 10.1177/0263276415590164.

Seigworth, G. J., & Gregg, M. (2010). An inventory of shimmers. In M. Gregg & G. J. Seigworth (Eds.), *The affect theory reader* (pp. 1–25). Durham, NC: Duke University Press.

Simmel, G. (2004). *The philosophy of money*. London: Routledge.

Wetherell, M. (2012). *Affect and emotion: A new social science understanding*. London: Sage.

Wetherell, M. (2014). Trends in the turn to affect: A social psychological critique. *Body and Society.* Online first. doi: 10.1177/1357034X14539020.

Index

© The Editor(s) (if applicable) and The Author(s) 2016
D. Beer, *Metric Power*, DOI 10.1057/978-1-137-55649-3